Doing Science

Adventures in Earth Science

Process-Oriented Activities for Grades 4–6

by Margy Kuntz

Fearon Teacher Aids
Carthage, Illinois

Simon & Schuster Supplementary Education Group

Thanks to Robert Cockerham from the U.S. Geological Survey in Palo Alto for the earthquake information on page 30.

Cover and interior illustrations by Bradley Dutsch

ISBN 0–8224–2318–9

Printed in the United States of America

Contents

Introduction

Teachers greatly influence children's interests. They expose children to new and stimulating topics and help them organize their knowledge. The *Doing Science* series was created to help you encourage children to be curious, to ask questions, to experiment, to learn, and to organize and integrate knowledge. This book will help you teach the processes of science—processes that can be integrated into all parts of our lives.

About This Book

Each **activity page** covers a topic that will easily fit into your science curriculum. The activities help students develop one or more of the following process skills:

- Observing—using the senses to gather information about objects and events.
- Comparing—identifying common and distinguishing characteristics among items or events.
- Measuring—comparatively or quantitatively describing the length, area, volume, mass, or temperature of objects.
- Classifying/Grouping—organizing information into logical categories.
- Sequencing—arranging items or events according to a characteristic.
- Collecting data—collecting and recording information obtained through observation.
- Organizing data—organizing data in a logical way so the results can be interpreted.
- Drawing conclusions—using the skills of inferring, predicting, and/or interpreting.

The **Teacher's Guide** will give you ideas for using each worksheet, including the main science concept, the process emphasis, and the materials list for each activity. The Teacher's Guide pages also include Discovery Questions—questions designed to make your students think and to encourage discussion. These questions are a mixture of specific-answer and open-ended questions that can be used either before, during, or after an activity. So while your students are doing science, they are also learning to think like scientists.

TEACHER'S GUIDE

Future Forecasts

Concept

Weather maps are useful sources of information.

Process Emphasis

Organizing data

Materials

For each student:

- Activity worksheet, page 24
- Pen or pencil

For the class:

- Weather maps from a local paper

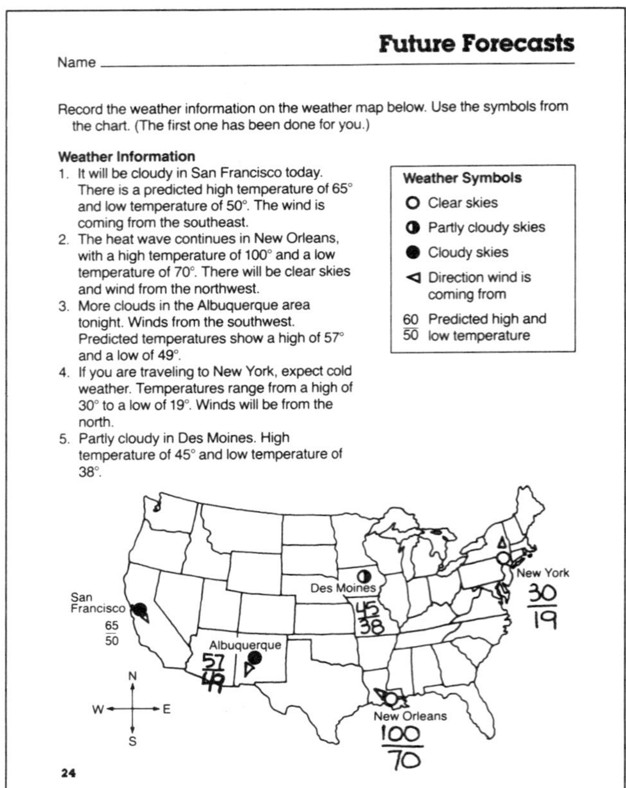

Future Forecasts

Name ______________________

Record the weather information on the weather map below. Use the symbols from the chart. (The first one has been done for you.)

Weather Information

1. It will be cloudy in San Francisco today. There is a predicted high temperature of 65° and low temperature of 50°. The wind is coming from the southeast.
2. The heat wave continues in New Orleans, with a high temperature of 100° and a low temperature of 70°. There will be clear skies and wind from the northwest.
3. More clouds in the Albuquerque area tonight. Winds from the southwest. Predicted temperatures show a high of 57° and a low of 49°.
4. If you are traveling to New York, expect cold weather. Temperatures range from a high of 30° to a low of 19°. Winds will be from the north.
5. Partly cloudy in Des Moines. High temperature of 45° and low temperature of 38°.

Weather Symbols

- O Clear skies
- ◑ Partly cloudy skies
- ● Cloudy skies
- ◁ Direction wind is coming from
- 60/50 Predicted high and low temperature

24

Procedure

1. Ask students to suggest types of information that should be included in a weather report. Be open to suggestions such as wind speed and direction, location of storm fronts, temperatures, and precipitation. Show students the weather maps. Explain that these maps show weather predictions for the area. Discuss the symbols on the maps. (Note: different areas use different symbols.)

2. Hand out the worksheet. Ask students to read the weather information and record it on the map. Ask them to use the symbols listed on the page. You might also have them record that day's prediction on their maps.

Discovery Questions

- What do you think a hurricane would look like on a weather map?
- What are some of the instruments people use to get information about weather?

Under Pressure

Concept

Barometers measure the air pressure and help us predict the weather.

Process Emphasis

Measuring and collecting data

Materials

For each student:

- Activity worksheet, page 25
- Pencil

For each group of four students:

- Large baby-food jar
- Scissors
- Balloon
- Paper or plastic straw
- Rubber bands
- White glue
- Cardboard
- Ruler

Procedure

1. Explain that a *barometer* measures air pressure. The needle on a barometer moves as the air pressure changes. Divide the class into groups of four. Hand out the materials and lead the class through the directions on the worksheet. Have students measure the barometric pressure twice a day, at the same times each day. They should also observe the weather changes during the week.

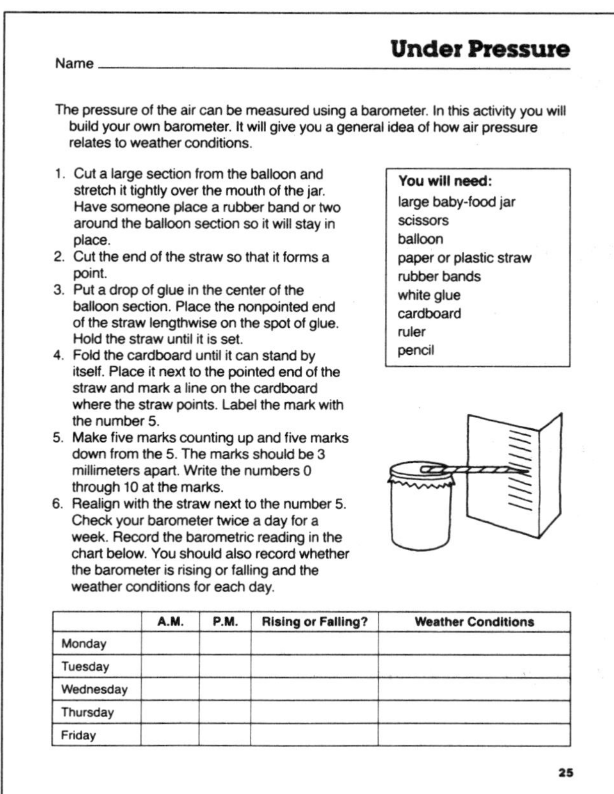

Under Pressure

Name ____________________

The pressure of the air can be measured using a barometer. In this activity you will build your own barometer. It will give you a general idea of how air pressure relates to weather conditions.

1. Cut a large section from the balloon and stretch it tightly over the mouth of the jar. Have someone place a rubber band or two around the balloon section so it will stay in place.
2. Cut the end of the straw so that it forms a point.
3. Put a drop of glue in the center of the balloon section. Place the nonpointed end of the straw lengthwise on the spot of glue. Hold the straw until it is set.
4. Fold the cardboard until it can stand by itself. Place it next to the pointed end of the straw and mark a line on the cardboard where the straw points. Label the mark with the number 5.
5. Make five marks counting up and five marks down from the 5. The marks should be 3 millimeters apart. Write the numbers 0 through 10 at the marks.
6. Realign with the straw next to the number 5. Check your barometer twice a day for a week. Record the barometric reading in the chart below. You should also record whether the barometer is rising or falling and the weather conditions for each day.

You will need:
large baby-food jar
scissors
balloon
paper or plastic straw
rubber bands
white glue
cardboard
ruler
pencil

	A.M.	P.M.	Rising or Falling?	Weather Conditions
Monday				
Tuesday				
Wednesday				
Thursday				
Friday				

25

2. At the end of the week, have students compare their barometric changes with the weather. Ask them to look for patterns, such as high readings bring clear weather and low readings precede storms.

Discovery Questions

- How could your barometer help you predict the weather?
- What is the difference between an aneroid barometer and a mercury barometer?

Hot Enough for You?

Concept

Sunlight affects air temperature.

Process Emphasis

Measuring, collecting data, and organizing data

Materials

For each student:

- Activity worksheet, page 26
- Colored pencils

For the class:

- Three thermometers

Hot Enough for You?

Name ____________________

Does the sun affect the temperature of the air? Try this experiment and decide for yourself.

1. Place the thermometers in three different places—one that gets direct sunlight, one that gets partial sunlight, and one that gets no sunlight.
2. Check the temperature on each thermometer at the same time each day. Record the temperatures in the chart below.
3. At the end of the week, plot the temperatures on the graph. Use a different color for each thermometer.

You will need:
Three thermometers

	Monday	Tuesday	Wednesday	Thursday	Friday
Direct sun					
Partial sun					
No sun					

Day of the Week: Monday, Tuesday, Wednesday, Thursday, Friday

20 30 40 50 60 70 80 90 100 110 120 130

Temperature (°)

- What was the highest temperature? ____________________
- Where was the thermometer with the highest temperature? ____________________

26

Procedure

1. Have students place the thermometers in three different places—one that gets direct sunlight, one that gets partial sunlight, and one that gets no sunlight. (Note: if you are doing this activity during the early fall or late spring, make sure your thermometers can handle very high temperatures.) Have students check the thermometers once a day for a week. They should check the temperatures at the same time each day and record their findings on the worksheet. At the end of the week, have students plot their findings on the graph provided. You may have to remind them how to plot information on graphs. Ask them to discuss their findings.

Discovery Questions

- What is the hottest place on earth?
- What is the coldest place on earth?
- What are some other factors besides sunlight that influence air temperature?

Blowing in the Wind

Concept

Wind speed can be observed.

Process Emphasis

Comparing and collecting data

Materials

For each student:

- Activity worksheet, page 27
- Pen or pencil

Name ______

Blowing in the Wind

You can estimate the speed of wind by using the Beaufort Scale. Observe objects outside once a day at the same time each day. Record your observations in the chart below. Then use the Beaufort Scale to find the name and speed of the wind. Write the information in the chart.

Beaufort Scale of Wind Speeds

Observation	Name of Wind	Miles per Hour
Smoke goes straight up	Calm	Less than 1
Smoke moves but weather vanes do not	Light Air	1–3
Weather vanes move; leaves rustle	Light Breeze	4–7
Flags flutter; leaves move constantly	Gentle Breeze	8–12
Dirt and paper raised; flags flap	Moderate Breeze	13–18
Small trees sway; flags ripple	Fresh Breeze	19–24
Large branches move; flags beat	Strong Breeze	25–31
Whole trees sway; flags are extended	Moderate Gale	32–38
Twigs break off; hard to walk against	Fresh Gale	39–46
Slight damage to buildings	Strong Gale	47–54
Trees uprooted; windows break	Full Gale	55–63
Widespread damage to buildings	Violent Storm	64–75
General destruction	Hurricane	Over 75

	Observations	Name of wind	Miles per hour
Monday			
Tuesday			
Wednesday			
Thursday			
Friday			

- Which day had the fastest wind? What was the wind speed? ______
- Which day had the slowest wind? What was the wind speed? ______

27

Procedure

1. Ask students to point out evidence that the air is moving (tree branches and grass bending, hair blowing, whistling sounds). Then ask them if they can think of a way to measure the speed of the wind without using any instruments. Explain that they could observe the motions of various objects such as flags, leaves, and chimney smoke to get an idea of how strong the wind is. Discuss the Beaufort Scale shown on the worksheet.

2. Hand out the worksheet. Have students observe the wind once a day for a week. (Measurements should be made at the same time each day.) Explain that they should record their observations and use the Beaufort Scale to determine the name and the strength of the wind.

Discovery Questions

- Which wind is stronger—a hurricane or a tornado?
- Which allows you to see very slight air movement—chimney smoke or large tree branches?
- Which allows you to see very strong air movement—chimney smoke or trees?

The Windy Cities

Concept

Wind speed can be measured.

Process Emphasis

Comparing and grouping

Materials

For each student:

- Activity worksheet, page 28
- Pen or pencil

Procedure

1. Ask students if they can think of a way to measure the wind. Draw a picture of an anemometer on the board or for an overhead projector (see Figure A). Explain that scientists measure wind speed with a device called an *anemometer*. Ask students if they can tell you how the anemometer works. (Cups on the device turn when wind hits them. The faster the wind, the faster the cups turn. The cups are connected to a dial that shows the wind speed.) Discuss their answers.

2. Hand out the worksheet. Explain that on weather maps, certain symbols are used to indicate wind speed. Discuss the symbols shown on the worksheet. Have students identify the wind speed for each city and write the names of the cities in the proper columns.

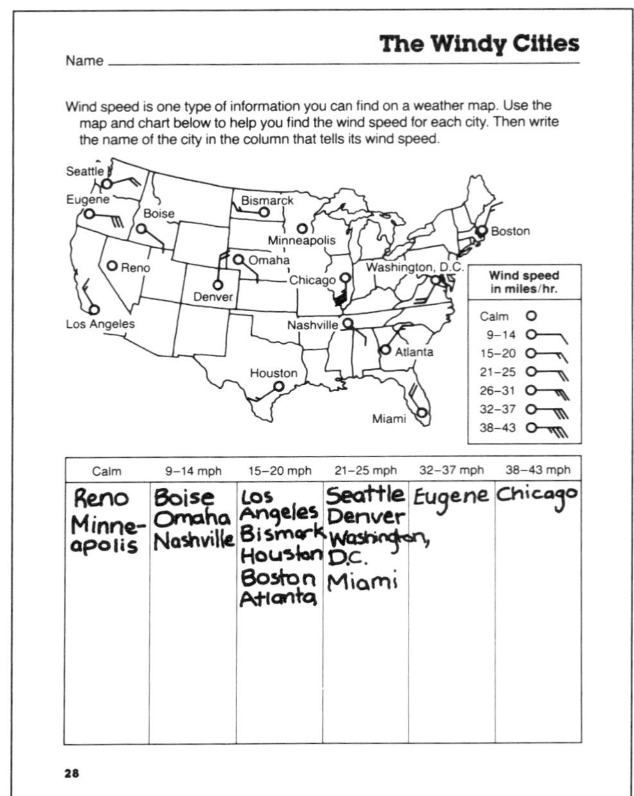

The Windy Cities

Name ______________________

Wind speed is one type of information you can find on a weather map. Use the map and chart below to help you find the wind speed for each city. Then write the name of the city in the column that tells its wind speed.

Calm	9–14 mph	15–20 mph	21–25 mph	32–37 mph	38–43 mph
Reno Minne-apolis	Boise Omaha Nashville	Los Angeles Bismark Houston Boston Atlanta	Seattle Denver Washington, D.C. Miami	Eugene	Chicago

28

Discovery Questions

- How is wind created?
- Does the wind always come from the same direction?

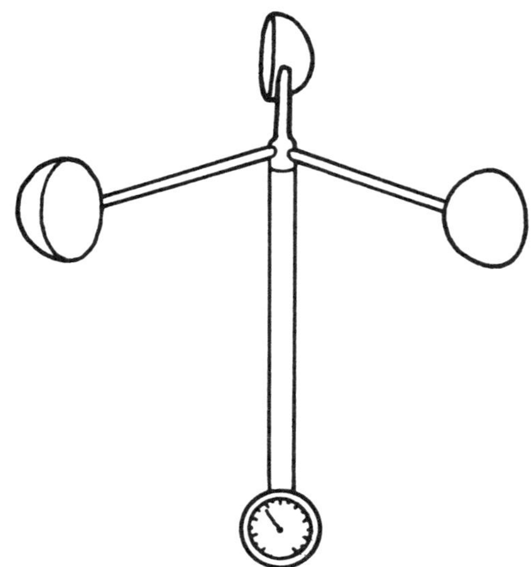

Figure A: anemometer

Here Today, Gone Tomorrow

Concept

Wind, water, and ice erosion change the shape of the land.

Process Emphasis

Sequencing

Materials

For each student:

- Activity worksheet, page 29
- Pen or pencil

For the class:

- A pile of dry sand
- A rectangular cake pan
- Ice cubes
- Large pitcher of water

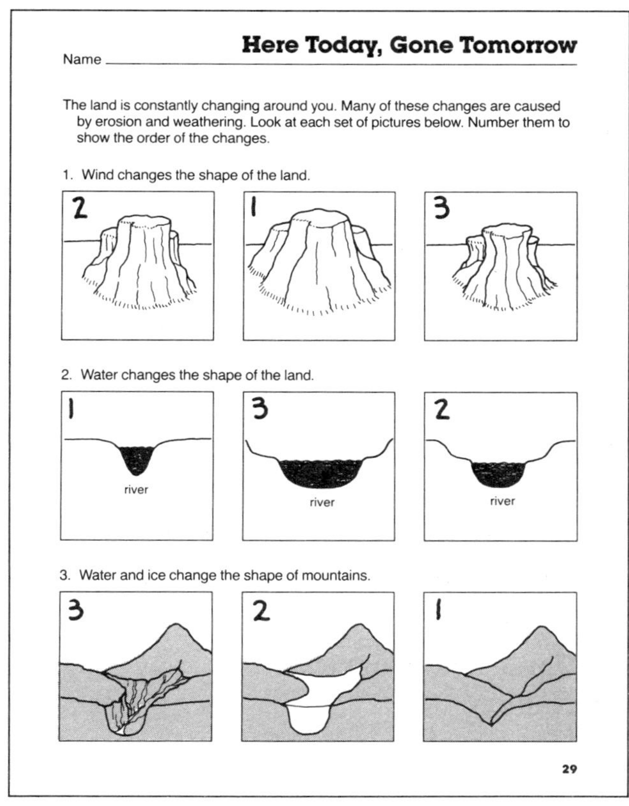

Here Today, Gone Tomorrow

Name ____________________

The land is constantly changing around you. Many of these changes are caused by erosion and weathering. Look at each set of pictures below. Number them to show the order of the changes.

1. Wind changes the shape of the land.

2   1   3

2. Water changes the shape of the land.

1   3   2

3. Water and ice change the shape of mountains.

3   2   1

29

Procedure

1. Demonstrate the effects of erosion in the following ways:
Wind Erosion: Pile the sand in the cake pan. Ask students what they think will happen if you blow on the sand. Blow on the sand. Have students observe and discuss the effects.
Ice Erosion: Pile wet sand in the cake pan. Ask students what they think will happen if you put ice cubes (to represent a glacier) on top of the sand mountain. Place the ice on top of the sand. Have students observe and discuss the effects.
Water Erosion: Pile wet sand in the cake pan. Make a narrow downhill trench in the sand to indicate a river's path. Ask students what they think will happen when water rushes through the sand. Pour a steady stream of water into the top of the trench. Have students observe and discuss the effects.

2. Hand out the worksheet. Have students number the pictures of erosion to show the order of the events.

Discovery Questions

- What is the difference between erosion and weathering?
- How was the Grand Canyon formed?

Shake, Rattle, and Roll

Concept

Earthquakes are a shaking of the earth caused by movement of rocks along a fault.

Process Emphasis

Organizing data and drawing conclusions

Materials

For each student:

- Activity worksheet, page 30
- Colored pencils

For the class:

- Three colors of clay
- Knife

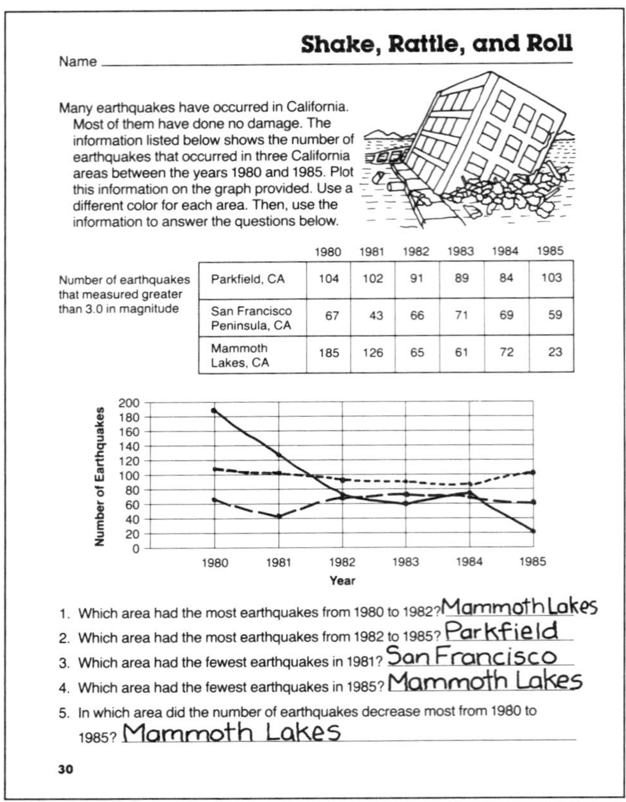

Shake, Rattle, and Roll

Name ____________________

Many earthquakes have occurred in California. Most of them have done no damage. The information listed below shows the number of earthquakes that occurred in three California areas between the years 1980 and 1985. Plot this information on the graph provided. Use a different color for each area. Then, use the information to answer the questions below.

Number of earthquakes that measured greater than 3.0 in magnitude		1980	1981	1982	1983	1984	1985
	Parkfield, CA	104	102	91	89	84	103
	San Francisco Peninsula, CA	67	43	66	71	69	59
	Mammoth Lakes, CA	185	126	65	61	72	23

1. Which area had the most earthquakes from 1980 to 1982? Mammoth Lakes
2. Which area had the most earthquakes from 1982 to 1985? Parkfield
3. Which area had the fewest earthquakes in 1981? San Francisco
4. Which area had the fewest earthquakes in 1985? Mammoth Lakes
5. In which area did the number of earthquakes decrease most from 1980 to 1985? Mammoth Lakes

30

Procedure

1. Flatten the pieces of clay. Tell the students that the clay represents layers of rocks in the earth's crust. Stack and press the layers on top of one another. Push in the sides so the clay folds up (see Figure A). Point out that just as the clay formed folds because of pressure, pressures from inside or outside the earth's crust can force layers of rock to fold up.

2. Explain to the class that a *fault* occurs when a break occurs in the earth's crust. The land can shift up or down along the crack or sideways along the crack. Slice the clay with a knife to separate it. Place the two pieces of clay near each other, but do not align them evenly. Tell the students that this is what a fault looks like. Explain that when huge rocks move along a fault, the ground shakes. This is called an earthquake. Discuss earthquakes and what happens during an earthquake.

3. Hand out the worksheet and pencils. Have the students plot the information on the graph. (You may want to demonstrate this graphing technique.) When the students have finished their graphs, they should answer the questions at the bottom of the page.

Discovery Questions

- What is the Richter Scale?
- Why do earthquakes often happen in California?

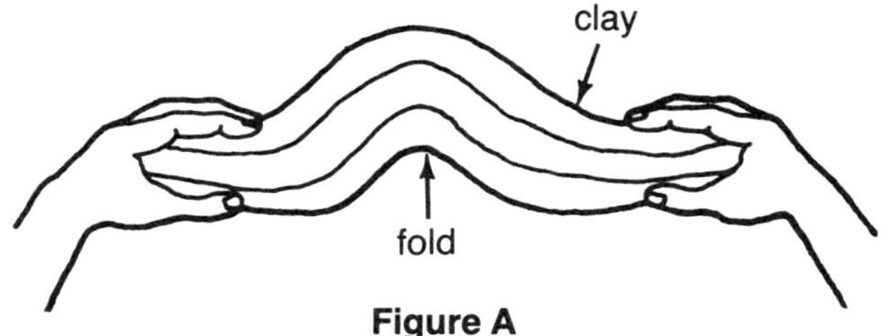

Figure A

Hot Rocks!

Concept

Igneous rocks are formed from hardened magma.

Process Emphasis

Classifying

Materials

For each student:

- Activity worksheet, page 31
- Pen or pencil
- Crayons or colored pencils (optional)

For the class:

- Samples of igneous rocks—granite, basalt, obsidian, and pumice
- Hand lenses (optional)

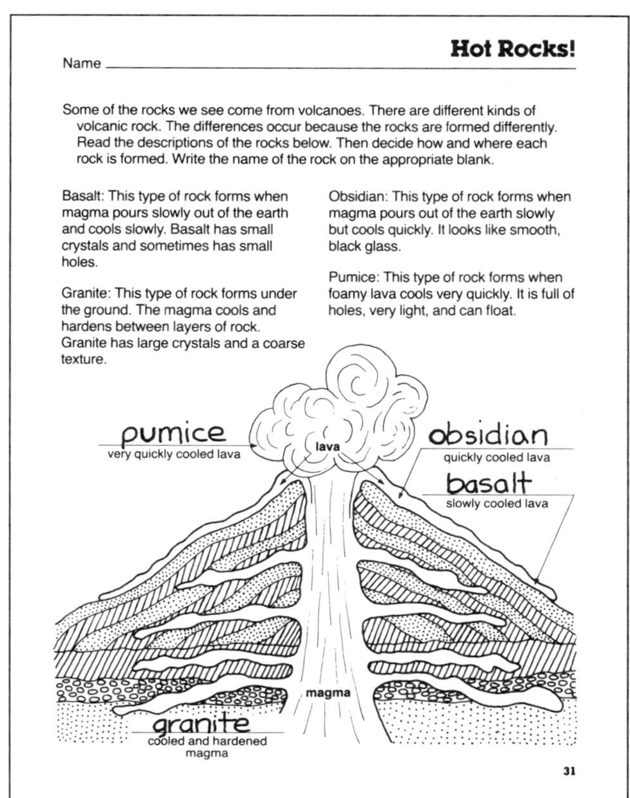

Hot Rocks!

Name ____________________

Some of the rocks we see come from volcanoes. There are different kinds of volcanic rock. The differences occur because the rocks are formed differently. Read the descriptions of the rocks below. Then decide how and where each rock is formed. Write the name of the rock on the appropriate blank.

Basalt: This type of rock forms when magma pours slowly out of the earth and cools slowly. Basalt has small crystals and sometimes has small holes.

Granite: This type of rock forms under the ground. The magma cools and hardens between layers of rock. Granite has large crystals and a coarse texture.

Obsidian: This type of rock forms when magma pours out of the earth slowly but cools quickly. It looks like smooth, black glass.

Pumice: This type of rock forms when foamy lava cools very quickly. It is full of holes, very light, and can float.

Procedure

1. Show students the rock samples. Explain that these rocks were formed from magma. (You also might want to explain that lava is magma found outside the earth.) Because of this some people call them "fire rocks." You might want to tell students that scientists call the rocks *igneous* rocks, which means rocks from fire. Have students study the samples (you might want to provide hand lenses) and compare their similarities and differences. Point out that the rocks are different because they were formed differently.

2. Hand out the worksheet. Have students write the name of each rock on the proper space. You might also have them color the different parts of the volcano.

Discovery Questions

- Gabbros is a type of rock that forms under the surface of the earth. The magma is slowly cooled and hardened. Does gabbros look more like granite or obsidian?
- Scoria is a rock that is formed by quickly cooled lava. Does scoria look more like granite or pumice?

Don't Take It for Granite!

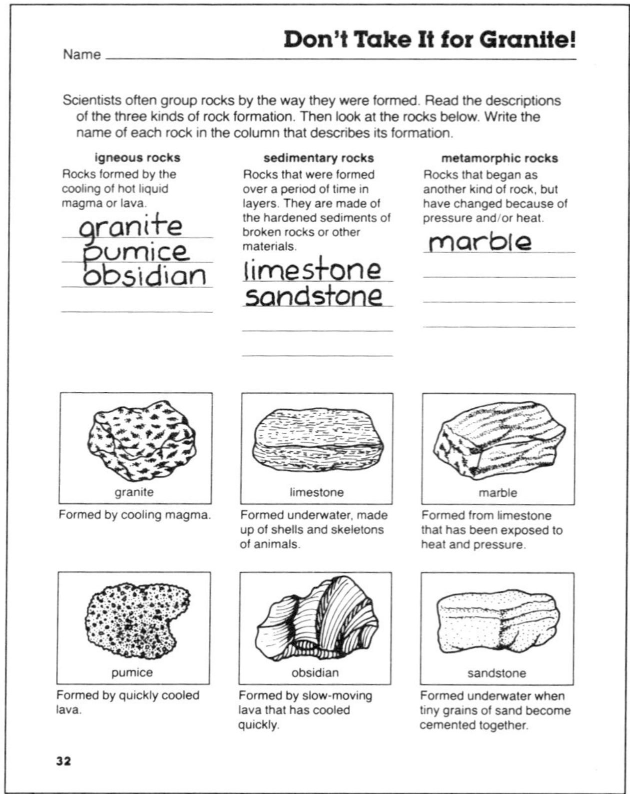

Don't Take It for Granite!

Name ______________________

Scientists often group rocks by the way they were formed. Read the descriptions of the three kinds of rock formation. Then look at the rocks below. Write the name of each rock in the column that describes its formation.

igneous rocks	sedimentary rocks	metamorphic rocks
Rocks formed by the cooling of hot liquid magma or lava.	Rocks that were formed over a period of time in layers. They are made of the hardened sediments of broken rocks or other materials.	Rocks that began as another kind of rock, but have changed because of pressure and/or heat.
granite	limestone	marble
pumice	sandstone	
obsidian		

granite
Formed by cooling magma.

limestone
Formed underwater, made up of shells and skeletons of animals.

marble
Formed from limestone that has been exposed to heat and pressure.

pumice
Formed by quickly cooled lava.

obsidian
Formed by slow-moving lava that has cooled quickly.

sandstone
Formed underwater when tiny grains of sand become cemented together.

32

Concept

Rocks form in different ways.

Process Emphasis

Grouping

Materials

For each student:

- Activity worksheet, page 32
- Pen or pencil

For the class:

- Samples of igneous rocks (granite, pumice, obsidian); sedimentary rocks (limestone, sandstone, chalk, clay); and metamorphic rocks (marble, slate, gneiss). These are usually available at lapidary stores.

Procedure

1. Show students the rock samples. Have students examine the rocks and describe the characteristics of each rock. Ask them to compare the similarities and the differences of the rocks and see if they can create a way to group them. Explain the different ways in which rocks form. Point out the examples of each form.

2. Hand out the worksheet. Have students read the descriptions of each rock and write the name of the rock in the appropriate blank.

Discovery Questions

- In which type of rock are you most likely to find fossils? Why?
- What can sedimentary rocks tell us about the past?

Scratch and Match

Concept

The minerals in some rocks can be identified by their color.

Process Emphasis

Observing, collecting data, and drawing conclusions

Materials

For each student:

- Activity worksheet, page 33
- Pen or pencil

For each group of four students:

- Piece of unglazed tile
- Four to eight of the following rocks: azurite, chalcopyrite or pyrite, cinnabar, galena*, hematite, limonite, magnetite, malachite or olivine. These are usually available at lapidary stores.

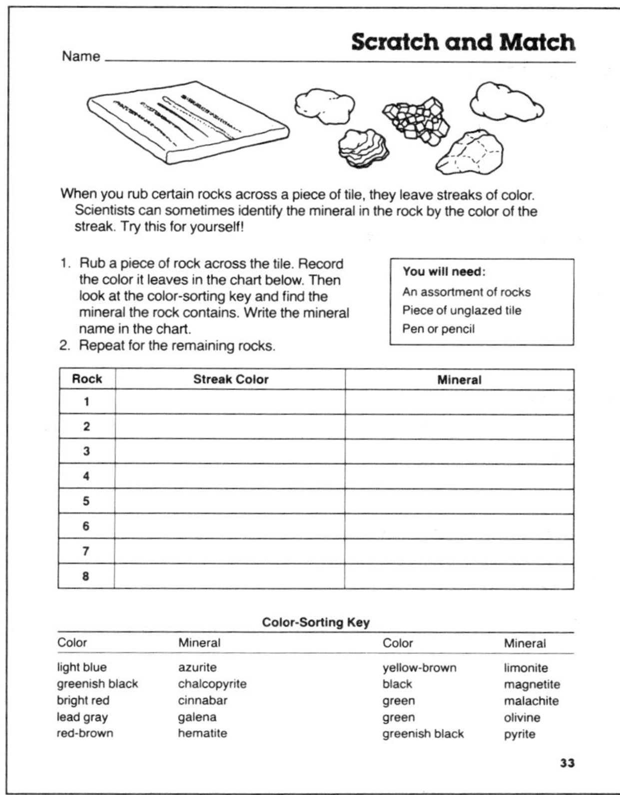

Scratch and Match

Name ______________________

When you rub certain rocks across a piece of tile, they leave streaks of color. Scientists can sometimes identify the mineral in the rock by the color of the streak. Try this for yourself!

1. Rub a piece of rock across the tile. Record the color it leaves in the chart below. Then look at the color-sorting key and find the mineral the rock contains. Write the mineral name in the chart.
2. Repeat for the remaining rocks.

You will need:
An assortment of rocks
Piece of unglazed tile
Pen or pencil

Rock	Streak Color	Mineral
1		
2		
3		
4		
5		
6		
7		
8		

Color-Sorting Key

Color	Mineral	Color	Mineral
light blue	azurite	yellow-brown	limonite
greenish black	chalcopyrite	black	magnetite
bright red	cinnabar	green	malachite
lead gray	galena	green	olivine
red-brown	hematite	greenish black	pyrite

33

Procedure

1. Divide the class into groups of four. Hand out the materials. Have the students rub one piece of rock at a time across the tile. (This will leave a streak of color on the tile.) Students should record the color of the streak on their worksheets. Have them use the color-sorting key on the page to identify the minerals in each rock.

2. Discuss the activity. Explain that all the rocks in the world are made of minerals—different rocks have different minerals and/or different amounts of minerals. You might ask students if they can think of other ways to identify the minerals in rocks.

Discovery Questions

- Is the streak test useful if you have several rocks that leave white streaks? Why?
- Why are there differences between the external color of a rock and the color of the streak it leaves?

*If you are using galena, make sure students wash their hands thoroughly after handling the material.

Telling Time with Fossils

Concept

Fossils are different ages.

Process Emphasis

Sequencing

Materials

For each student:

- Activity worksheet, page 34
- Pen or pencil

For the class:

- Picture or diagram of sedimentary rock layers

Telling Time with Fossils

Name ______________________

Scientists can "see back in time" using fossils. Old fossils tell us something about the earth's past. Some fossils are more than two billion years old. Create a fossil time line. Look at the pictures below. Then write the name of each creature under its age on the time line.

Cephalaspis ≈ 410 million years old	Eohippus ≈ 50 million years old	Australopithecus ≈ 2 million years old	Trilobite ≈ 600 million years old
Dimetrodon ≈ 275 million years old	**Stegasaurus** ≈ 150 million years old	**Ichthyostega** ≈ 365 million years old	**Diatryma** ≈ 65 million years old

Millions of years ago

600 500 400 300 200 100 50 0

410 365 275 150 65 2

Trilobite, Cephalaspis, Ichthyostega, Dimetrodon, Stegasaurus, Diatryma, Eohippus, Australopithecus

34

Procedure

1. Discuss how sedimentary rock forms. Show students the picture of the rock layers. Explain that scientists find different fossils in the different layers of rock. Since the rock layers build up over time, the oldest layers are on the bottom and the younger layers are on the top. Point out that the oldest fossils are found in the lowest layers of rock. If an outside area is available for digging, you might let students dig in the soil to see if they can detect a difference in layers.

2. Hand out the worksheet. Have students write the name of the organism under its age on the time line.

Discovery Questions

- How are fossils formed?
- How can scientists tell if organisms have changed over time?

Fossil Circles

Concept

Fossils can be classified in many ways.

Process Emphasis

Grouping

Materials

For each student:

- Activity worksheets, pages 35 and 36
- Scissors
- Four pieces 24″ yarn; three yellow and one red

Procedure

1. Duplicate the worksheets on heavy paper and hand out. Have students color the labels on page 35 according to the directions. They should cut out the cards and labels on both pages.

2. Ask students to tie each piece of yarn into a loop. They should have three yellow loops and one red loop. Have students lay the red loop on their desks in the shape of a circle. Then have them overlap the red circle with the three yellow circles, so the arrangement looks like this:

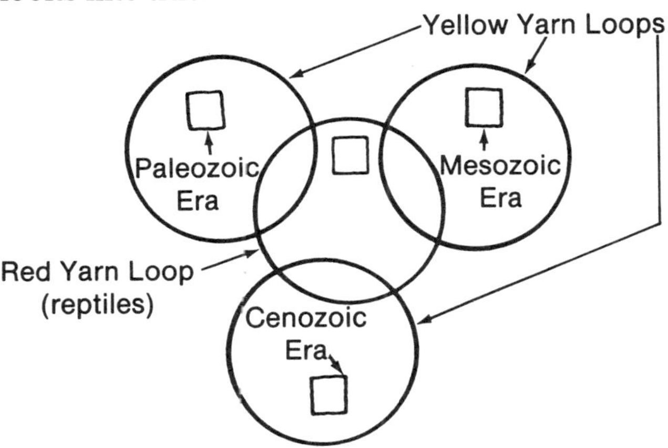

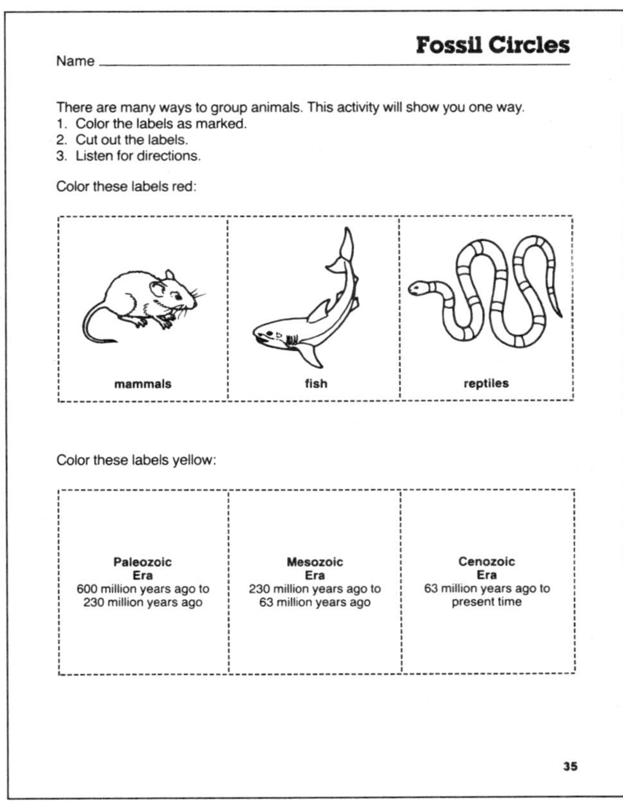

Fossil Circles

Name ______

There are many ways to group animals. This activity will show you one way.
1. Color the labels as marked.
2. Cut out the labels.
3. Listen for directions.

Color these labels red:

mammals	fish	reptiles

Color these labels yellow:

Paleozoic Era 600 million years ago to 230 million years ago	Mesozoic Era 230 million years ago to 63 million years ago	Cenozoic Era 63 million years ago to present time

35

Fossil Circles

Name ______

You will use these cards with the labels on page 35.
1. Cut out the cards.
2. Listen for directions.

Paleozoic Era	Paleozoic Era	Cenozoic Era	Mesozoic Era	Cenozoic Era
Mesozoic Era	Cenozoic Era	Cenozoic Era	Mesozoic Era	Paleozoic Era
Cenozoic Era	Mesozoic Era	Paleozoic Era	Paleozoic Era	Cenozoic Era
Mesozoic Era	Mesozoic Era	Cenozoic Era	Mesozoic Era	Mesozoic Era

36

The red label *reptiles* should be put inside the red loop, and the three yellow labels *Paleozoic Era, Mesozoic Era,* and *Cenozoic Era* should be placed inside the yellow circles. Make sure the students put the yellow labels outside the intersection points of the yellow and red circles.

3. Have students group the fossil cards from page 36 by age. (Note: the ages of the fossils are on the cards.) Then have them put the cards in the appropriate yellow circles. Ask students to find the fossils in each circle that are reptiles. Have them move these cards to the intersection spaces between the yellow and red loops. When they have finished the exercise you might have them explain why we would want to organize the cards in this fashion.

4. Repeat the process using one of the other red labels.

Discovery Questions

- To which era do human beings belong?
- Why did the dinosaurs become extinct?

Inside or Out?

Concept

We can group the nine planets of our Solar System into two groups—inner planets (the four planets closest to the sun) and outer planets (the five planets farthest from the sun).

Process Emphasis

Classifying

Materials

For each student:

- Activity worksheet, page 37
- Pen or pencil

For the class:

- Pictures of the nine planets

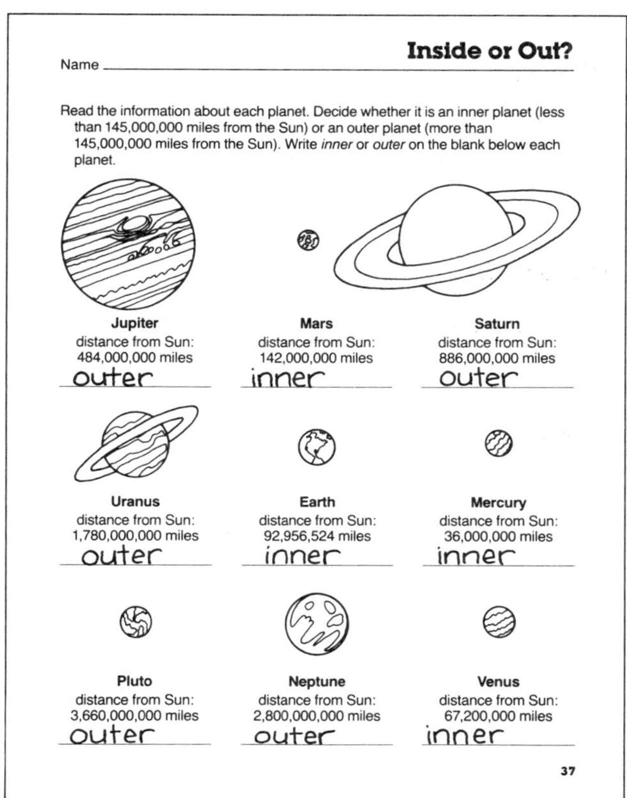

Name ______ **Inside or Out?**

Read the information about each planet. Decide whether it is an inner planet (less than 145,000,000 miles from the Sun) or an outer planet (more than 145,000,000 miles from the Sun). Write *inner* or *outer* on the blank below each planet.

Jupiter distance from Sun: 484,000,000 miles outer	**Mars** distance from Sun: 142,000,000 miles inner	**Saturn** distance from Sun: 886,000,000 miles outer
Uranus distance from Sun: 1,780,000,000 miles outer	**Earth** distance from Sun: 92,956,524 miles inner	**Mercury** distance from Sun: 36,000,000 miles inner
Pluto distance from Sun: 3,660,000,000 miles outer	**Neptune** distance from Sun: 2,800,000,000 miles outer	**Venus** distance from Sun: 67,200,000 miles inner

37

Procedure

1. Show students the pictures of the planets. Ask them to tell you as much as they can about each planet. Discuss the similarities and differences between the planets. Explain that there are many ways to compare the planets and one way is to compare their distances from the sun.

2. Hand out the worksheet. Have students classify each planet as an inner planet or an outer planet. They should write *inner* or *outer* on the blank below each planet.

Discovery Questions

- What are some of the similarities the four innermost planets share?
- What are some of the similarities the four largest planets share?
- How is Pluto similar to Earth?
- How is Pluto different from Earth?

As the Worlds Turn

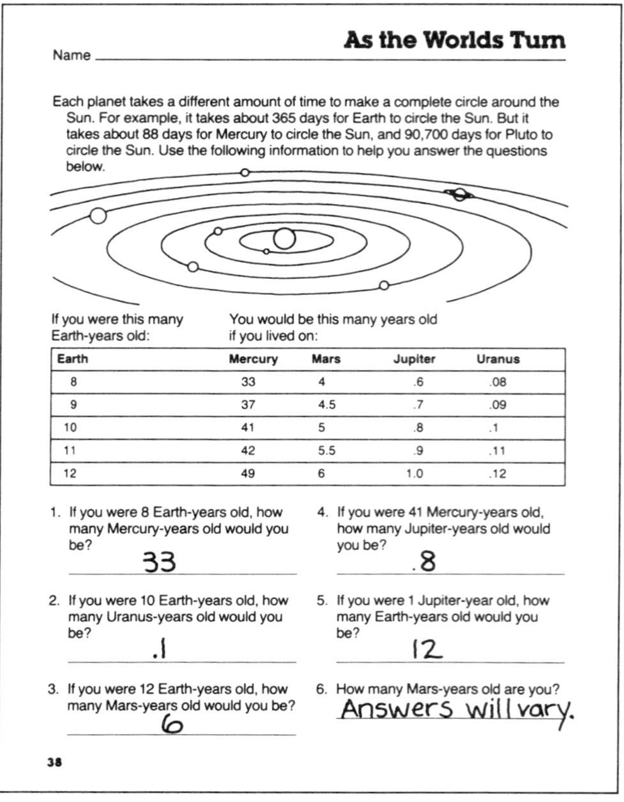

As the Worlds Turn

Name ____________________

Each planet takes a different amount of time to make a complete circle around the Sun. For example, it takes about 365 days for Earth to circle the Sun. But it takes about 88 days for Mercury to circle the Sun, and 90,700 days for Pluto to circle the Sun. Use the following information to help you answer the questions below.

If you were this many Earth-years old: You would be this many years old if you lived on:

Earth	Mercury	Mars	Jupiter	Uranus
8	33	4	.6	.08
9	37	4.5	.7	.09
10	41	5	.8	.1
11	42	5.5	.9	.11
12	49	6	1.0	.12

1. If you were 8 Earth-years old, how many Mercury-years old would you be? 33
2. If you were 10 Earth-years old, how many Uranus-years old would you be? .1
3. If you were 12 Earth-years old, how many Mars-years old would you be? 6
4. If you were 41 Mercury-years old, how many Jupiter-years old would you be? .8
5. If you were 1 Jupiter-year old, how many Earth-years old would you be? 12
6. How many Mars-years old are you? Answers will vary.

38

Concept

The planets take different amounts of time to circle around the Sun.

Process Emphasis

Comparing and drawing conclusions

Materials

For each student:

- Activity worksheet, page 38
- Pen or pencil

Procedure

1. Discuss the term *revolution*. Explain that it takes about 365 days for Earth to make a complete circle around the Sun. (You might want to demonstrate revolution and rotation by having students act as the Sun and Earth.) Tell students that the other planets in the solar system also revolve around the Sun. However, it takes each planet a different amount of time to do so.

2. Hand out the worksheet. Tell students that the chart on the worksheet shows equivalent number of years. For example, it takes Mercury about 88 days to circle the Sun. Therefore, in the time it takes Earth to revolve around the Sun one time, Mercury has revolved around the Sun about four times. Another way to say this is one Earth year is the same amount of time as four Mercury years. Have students answer the questions on the page using the information from the chart.

Discovery Questions

- If you were eleven years old on Earth, would you be older or younger if you lived on Mercury?
- Why do some planets take longer to revolve around the Sun than others?

Faces of the Moon

Concept

As the moon revolves, we can see different amounts of the lighted surface.

Process Emphasis

Comparing and drawing conclusions

Materials

For each student:

- Activity worksheet, page 39
- Pencil

For the class:

- A dark rubber ball
- Flashlight

Faces of the Moon

Name ____________________

The pictures below show the positions of the moon over a period of a month. The names of the phases we see on earth are written next to each circle. In each circle, draw what the moon would look like from earth during that phase. (The first two are done for you.)

Part You See
New Moon (invisible)
Waxing Crescent
Waning Crescent
First Quarter
Earth
Third Quarter
Waxing Gibbous
Waning Gibbous
Full Moon

39

Procedure

1. Darken the room. Aim the flashlight beam toward the front of the room. Hold the ball in the light beam and have students observe the patterns of light and dark on the ball from different areas of the room. When all students have had the opportunity to observe the patterns, slowly pass the ball through the light. Point out that the light patterns on the ball are similar to the way the moon looks when you observe it for several weeks. Discuss the phases of the moon and the names of the phases.

2. Hand out the worksheet. Explain that the diagram on the page shows the positions of the moon during a month. Students should draw each phase as we would see it on earth.

Discovery Questions

- Does the motion of the moon have an effect on the earth? If so, what is the effect?
- What did we learn about the moon from the Apollo landings?

Eclipse It!

Concept

Eclipses of the sun and moon depend on the positions and movements of the earth, sun, and moon.

Process Emphasis

Comparing

Materials

For each student:

- Activity worksheet, page 40
- Pen or pencil

For the class:

- Gooseneck lamp without a shade
- Globe of the earth
- Small ball
- Small soda bottle

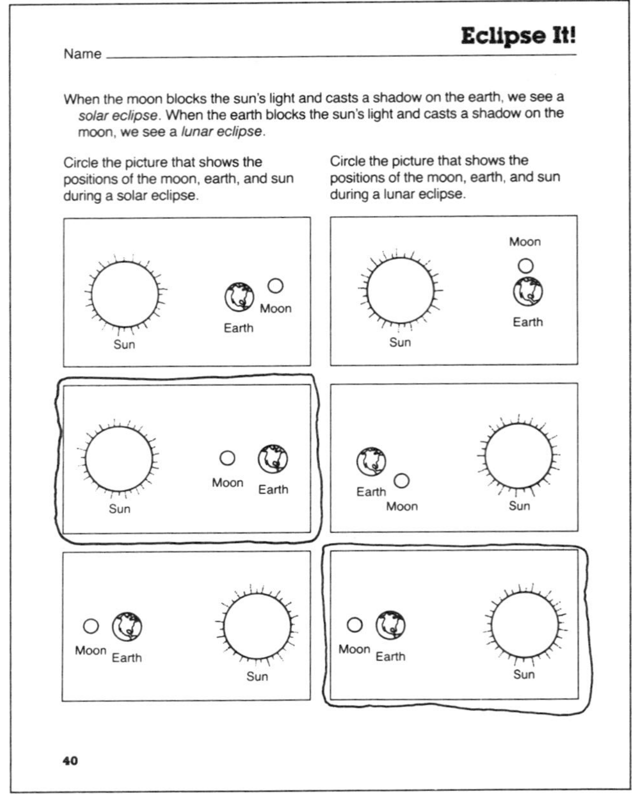

Eclipse It!

Name ____________

When the moon blocks the sun's light and casts a shadow on the earth, we see a *solar eclipse*. When the earth blocks the sun's light and casts a shadow on the moon, we see a *lunar eclipse*.

Circle the picture that shows the positions of the moon, earth, and sun during a solar eclipse.

Circle the picture that shows the positions of the moon, earth, and sun during a lunar eclipse.

40

Procedure

1. Ask students if they have ever witnessed a solar or lunar eclipse. Explain that eclipses occur when the earth and moon get in the way of each other's shadows. Demonstrate an eclipse using the lamp as the sun, the globe as the earth, and the small ball as the moon. Place the small ball on the top of the soda bottle and position it between the lamp and the globe so that a shadow of the ball is cast on the globe. Explain that this represents a solar eclipse. Point out that the distance and position of the moon relative to the earth are important. If the moon is too far away from the earth or is to the side of the earth, it will not completely block the sun's light.

2. Hand out the worksheet. Have students compare the drawings and choose the one that shows the proper positions for a lunar eclipse and a solar eclipse.

Discovery Questions

- What is an annular eclipse?
- Could you see solar eclipses if you were on another planet?

The Constellation Game

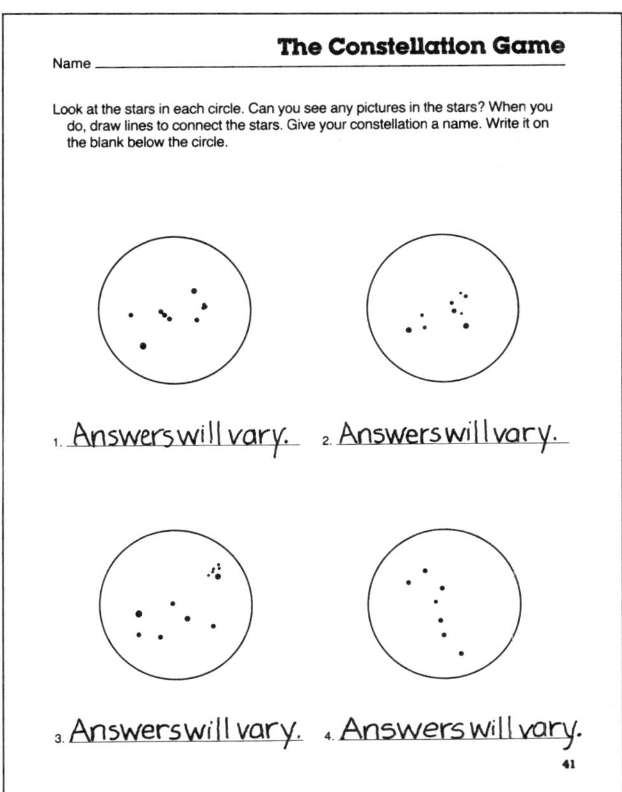
The Constellation Game

Name

Look at the stars in each circle. Can you see any pictures in the stars? When you do, draw lines to connect the stars. Give your constellation a name. Write it on the blank below the circle.

1. Answers will vary. 2. Answers will vary.

3. Answers will vary. 4. Answers will vary.

41

Concept

People see different patterns in groups of stars.

Process Emphasis

Grouping

Materials

For each student:

- Activity worksheet, page 41
- Pencil

Procedure

1. Explain that years ago people imagined that groups of stars looked like objects or creatures. They would name the groups of stars (constellations) after the creatures or objects they resembled.

2. Hand out the worksheet. Tell students that they should create their own constellations using the groups of stars shown. Have them connect the stars by drawing lines between them. Ask students to name their constellations.

3. When the class is finished, have students display their groupings and compare them with those of their classmates. Point out that different people saw different things in the stars. Explain that this used to happen all the time until a group of people (the International Astronomical Union) decided on set names for the constellations. Show students the standard names and patterns for the constellations on the worksheet. (1. Orion 2. Leo 3. Taurus 4. Big Dipper)

Discovery Questions

- Why did people see pictures in the stars?
- Are the stars in the constellations really right next to each other?

Starry, Starry Sight

Concept

There are recognizable constellations in the sky.

Process Emphasis

Observing and grouping

Materials

For each student:

- Activity worksheet, page 42
- Pencil

Procedure

1. Draw dots on the chalkboard to represent the stars of the Big Dipper (see Figure A). Ask students if they recognize the pattern on the board. Connect the dots with a solid line to show the pattern. Ask students if this pattern is easier or harder to see once the lines are drawn. Explain that when we look at constellations in the sky we have to imagine the lines between the stars. Sometimes it is easier to find a constellation if you look for a small part of it (for instance, the cup of the Big Dipper).

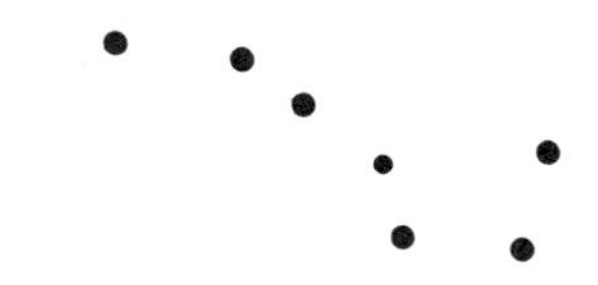

Figure A: Stars in the Big Dipper

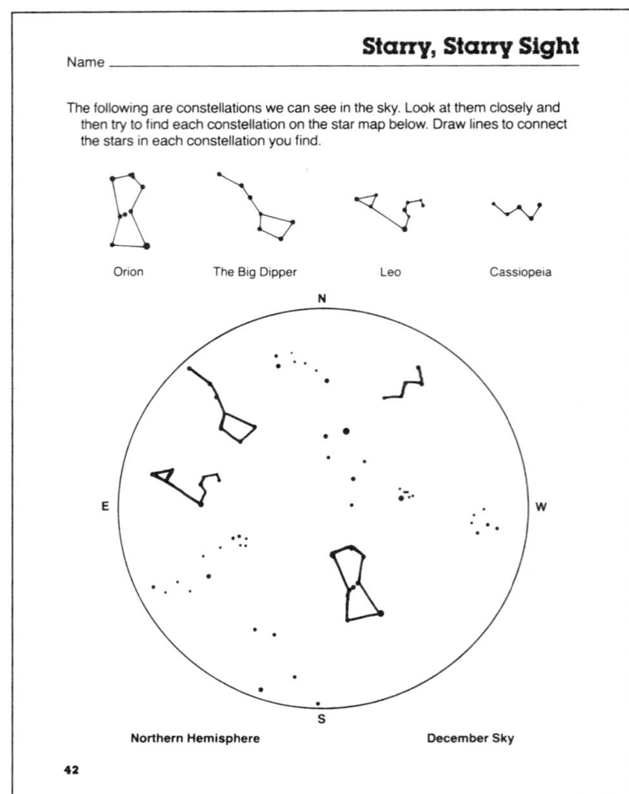

Name ____________ **Starry, Starry Sight**

The following are constellations we can see in the sky. Look at them closely and then try to find each constellation on the star map below. Draw lines to connect the stars in each constellation you find.

Northern Hemisphere December Sky

42

2. Hand out the worksheet. Have students find the constellations on the star map. Ask them to connect the stars once they have found the constellations.

Discovery Questions

- Would you see the same stars in the sky in January and July? Why?
- Do people in the United States see the same stars as people in England? Why?
- Do people in the United States see the same stars as people in Australia? Why?

How Hot Is Red?

Concept

Stars can be classified by their color and their temperature.

Process Emphasis

Organizing data and drawing conclusions

Materials

For each student:

- Activity worksheet, page 43
- Pencil or colored pencils

For the class:

- Matches, bunsen burner, or candle (optional)

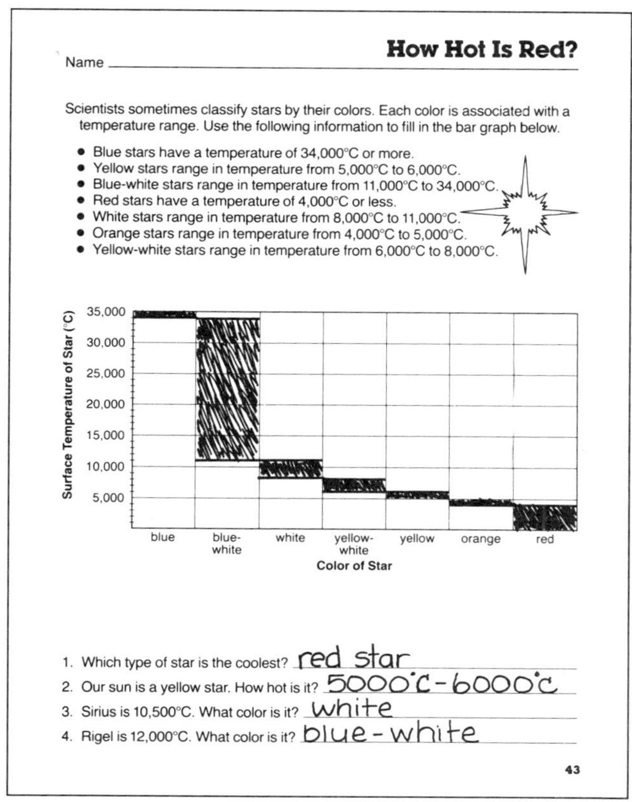

Name ______ **How Hot Is Red?**

Scientists sometimes classify stars by their colors. Each color is associated with a temperature range. Use the following information to fill in the bar graph below.

- Blue stars have a temperature of 34,000°C or more.
- Yellow stars range in temperature from 5,000°C to 6,000°C.
- Blue-white stars range in temperature from 11,000°C to 34,000°C.
- Red stars have a temperature of 4,000°C or less.
- White stars range in temperature from 8,000°C to 11,000°C.
- Orange stars range in temperature from 4,000°C to 5,000°C.
- Yellow-white stars range in temperature from 6,000°C to 8,000°C.

1. Which type of star is the coolest? red star
2. Our sun is a yellow star. How hot is it? 5000°C-6000°C
3. Sirius is 10,500°C. What color is it? white
4. Rigel is 12,000°C. What color is it? blue-white

43

Procedure

1. Ask students what color stars are. (Most students will probably say stars are white.) Explain that stars are actually different colors—blue, white, yellow, orange, or red. Discuss the fact that stars differ in color because they have different surface temperatures. You might want to demonstrate that color is related to temperature using matches, a bunsen burner, or a candle. (If you have the materials, heating a piece of iron will show the range of color from red, cool, to blue-white, very hot.)

2. Hand out the worksheet. Explain to students that they will be graphing *ranges* of temperature. You may want to demonstrate this graphing technique. When the students have finished their graphs, they should answer the questions at the bottom of the page.

Discovery Questions

- Why are stars hot?
- Is the temperature of the inside of a star different from the temperature on the surface of a star? If so, why?

ACTIVITY WORKSHEETS

Future Forecasts

Name ____________________

Record the weather information on the weather map below. Use the symbols from the chart. (The first one has been done for you.)

Weather Information

1. It will be cloudy in San Francisco today. There is a predicted high temperature of 65° and low temperature of 50°. The wind is coming from the southeast.
2. The heat wave continues in New Orleans, with a high temperature of 100° and a low temperature of 70°. There will be clear skies and wind from the northwest.
3. More clouds in the Albuquerque area tonight. Winds from the southwest. Predicted temperatures show a high of 57° and a low of 49°.
4. If you are traveling to New York, expect cold weather. Temperatures range from a high of 30° to a low of 19°. Winds will be from the north.
5. Partly cloudy in Des Moines. High temperature of 45° and low temperature of 38°.

Weather Symbols

- O Clear skies
- ◑ Partly cloudy skies
- ● Cloudy skies
- ◁ Direction wind is coming from
- $\frac{60}{50}$ Predicted high and low temperature

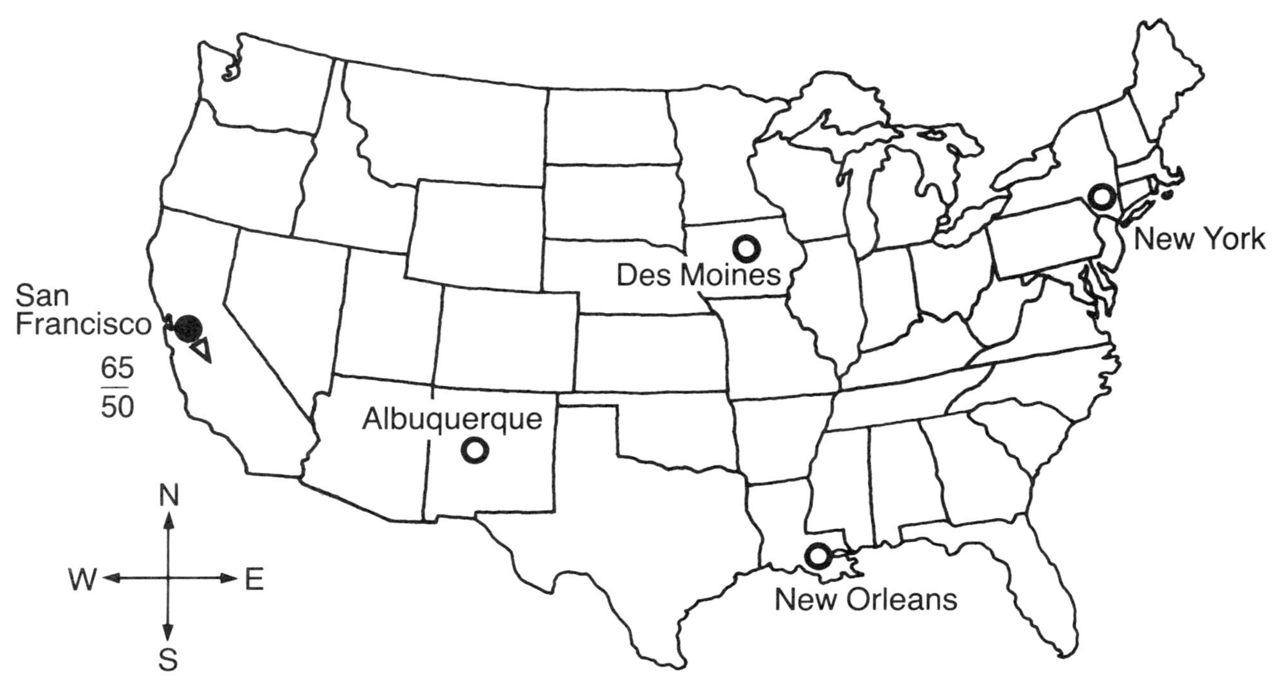

Under Pressure

Name ______________________________

The pressure of the air can be measured using a barometer. In this activity you will build your own barometer. It will give you a general idea of how air pressure relates to weather conditions.

You will need:
large baby-food jar
scissors
balloon
paper or plastic straw
rubber bands
white glue
cardboard
ruler
pencil

1. Cut a large section from the balloon and stretch it tightly over the mouth of the jar. Have someone place a rubber band or two around the balloon section so it will stay in place.
2. Cut the end of the straw so that it forms a point.
3. Put a drop of glue in the center of the balloon section. Place the nonpointed end of the straw lengthwise on the spot of glue. Hold the straw until it is set.
4. Fold the cardboard until it can stand by itself. Place it next to the pointed end of the straw and mark a line on the cardboard where the straw points. Label the mark with the number 5.
5. Make five marks counting up and five marks down from the 5. The marks should be 3 millimeters apart. Write the numbers 0 through 10 at the marks.
6. Realign with the straw next to the number 5. Check your barometer twice a day for a week. Record the barometric reading in the chart below. You should also record whether the barometer is rising or falling and the weather conditions for each day.

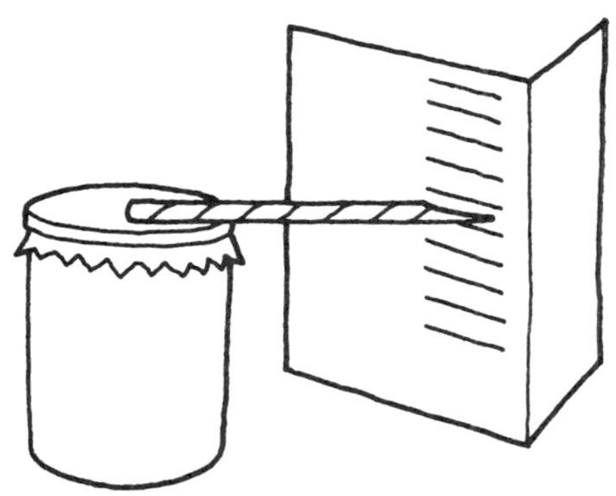

	A.M.	P.M.	Rising or Falling?	Weather Conditions
Monday				
Tuesday				
Wednesday				
Thursday				
Friday				

Hot Enough for You?

Name ______________________________

Does the sun affect the temperature of the air? Try this experiment and decide for yourself.

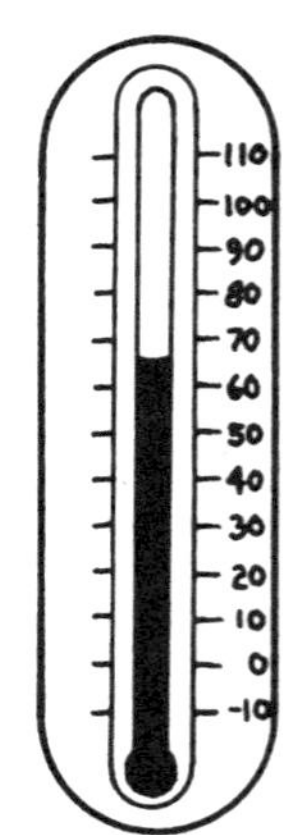

1. Place the thermometers in three different places—one that gets direct sunlight, one that gets partial sunlight, and one that gets no sunlight.
2. Check the temperature on each thermometer at the same time each day. Record the temperatures in the chart below.
3. At the end of the week, plot the temperatures on the graph. Use a different color for each thermometer.

You will need:

Three thermometers

	Monday	Tuesday	Wednesday	Thursday	Friday
Direct sun					
Partial sun					
No sun					

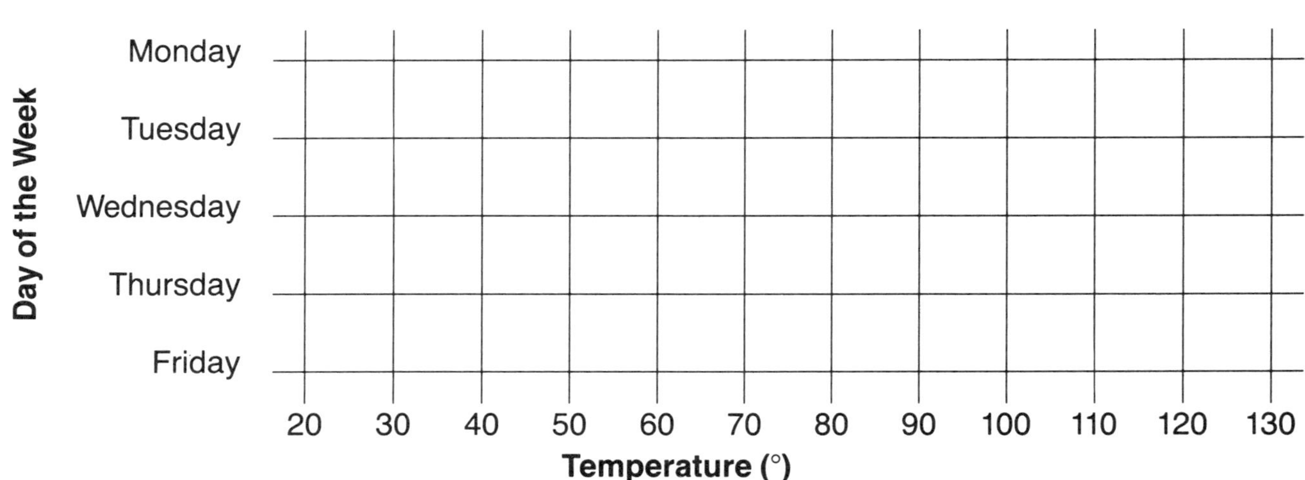

- What was the highest temperature? ______________________________
- Where was the thermometer with the highest temperature? ______________________________

Blowing in the Wind

Name ______________________________

You can estimate the speed of wind by using the Beaufort Scale. Observe objects outside once a day at the same time each day. Record your observations in the chart below. Then use the Beaufort Scale to find the name and speed of the wind. Write the information in the chart.

Beaufort Scale of Wind Speeds

Observation	**Name of Wind**	**Miles per Hour**
Smoke goes straight up	Calm	Less than 1
Smoke moves but weather vanes do not	Light Air	1–3
Weather vanes move; leaves rustle	Light Breeze	4–7
Flags flutter; leaves move constantly	Gentle Breeze	8–12
Dirt and paper raised; flags flap	Moderate Breeze	13–18
Small trees sway; flags ripple	Fresh Breeze	19–24
Large branches move; flags beat	Strong Breeze	25–31
Whole trees sway; flags are extended	Moderate Gale	32–38
Twigs break off; hard to walk against	Fresh Gale	39–46
Slight damage to buildings	Strong Gale	47–54
Trees uprooted; windows break	Full Gale	55–63
Widespread damage to buildings	Violent Storm	64–75
General destruction	Hurricane	Over 75

	Observations	**Name of wind**	**Miles per hour**
Monday			
Tuesday			
Wednesday			
Thursday			
Friday			

- Which day had the fastest wind? What was the wind speed? ____________
- Which day had the slowest wind? What was the wind speed? ____________

The Windy Cities

Name ______________________________

Wind speed is one type of information you can find on a weather map. Use the map and chart below to help you find the wind speed for each city. Then write the name of the city in the column that tells its wind speed.

Calm	9–14 mph	15–20 mph	21–25 mph	32–37 mph	38–43 mph

Here Today, Gone Tomorrow

Name ___________________________

The land is constantly changing around you. Many of these changes are caused by erosion and weathering. Look at each set of pictures below. Number them to show the order of the changes.

1. Wind changes the shape of the land.

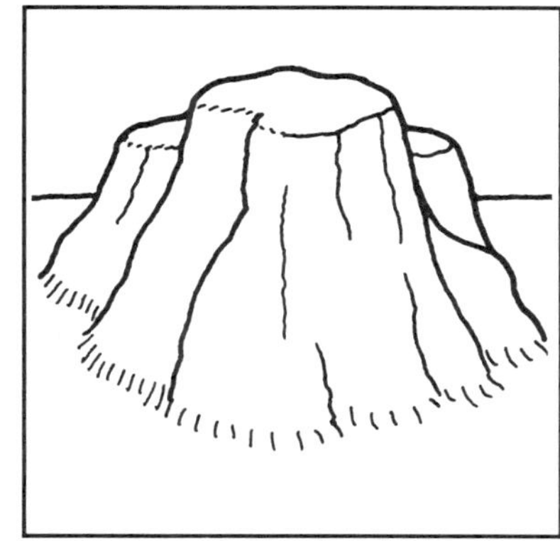
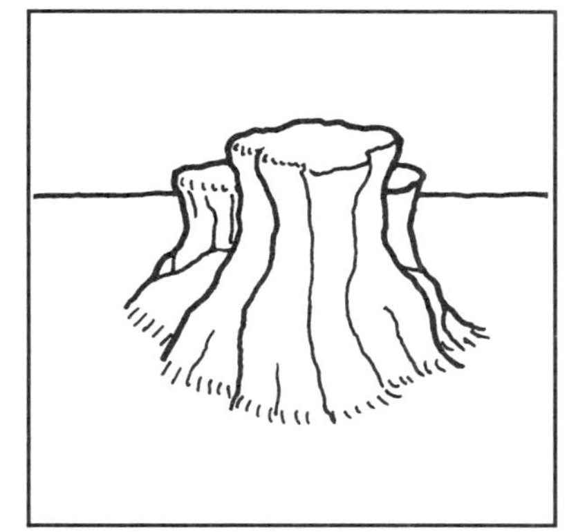

2. Water changes the shape of the land.

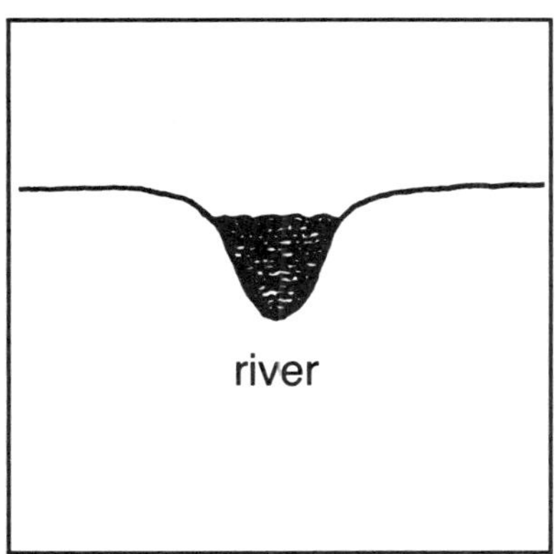

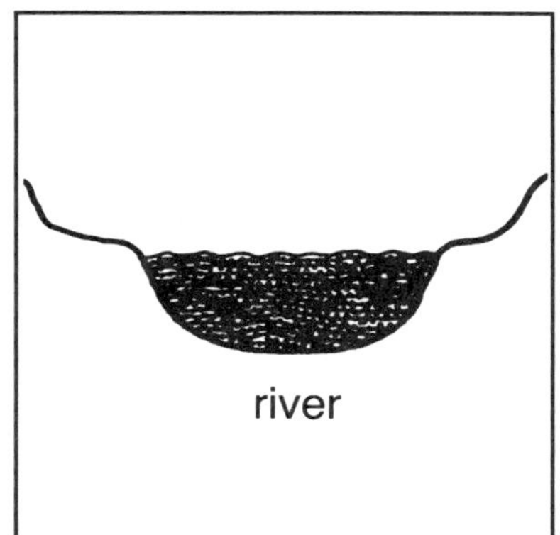

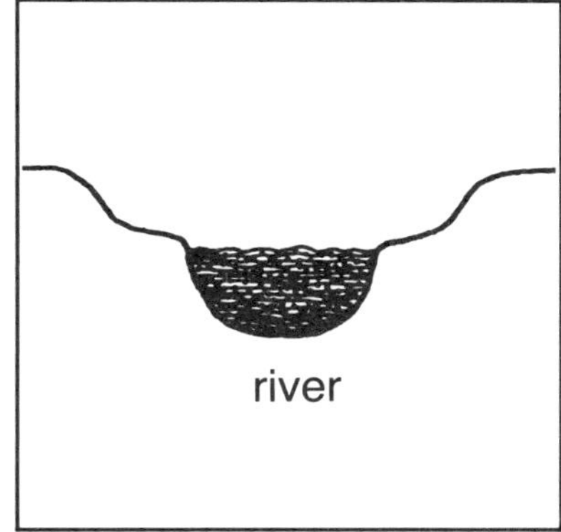

3. Water and ice change the shape of mountains.

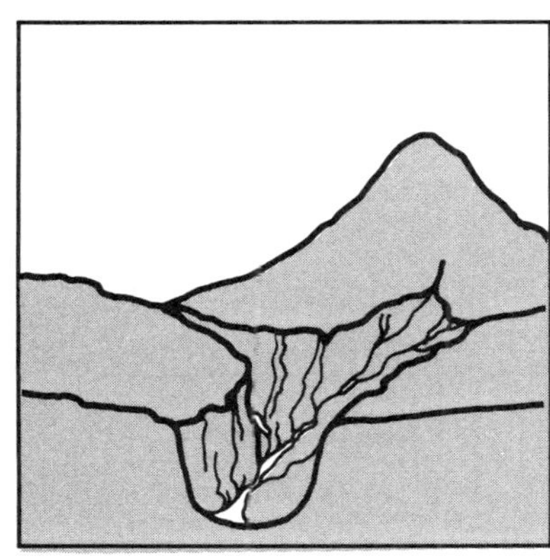
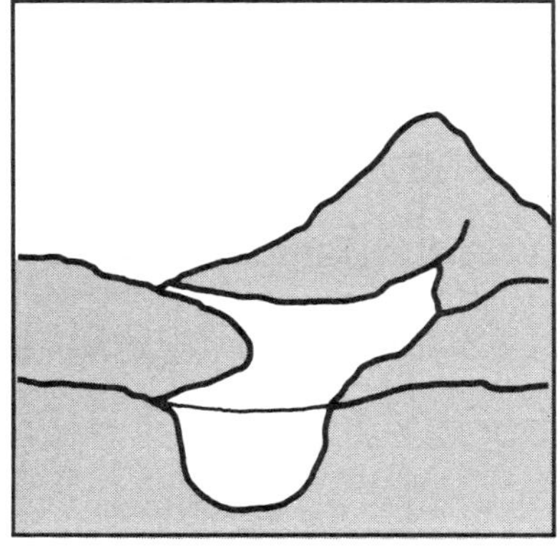
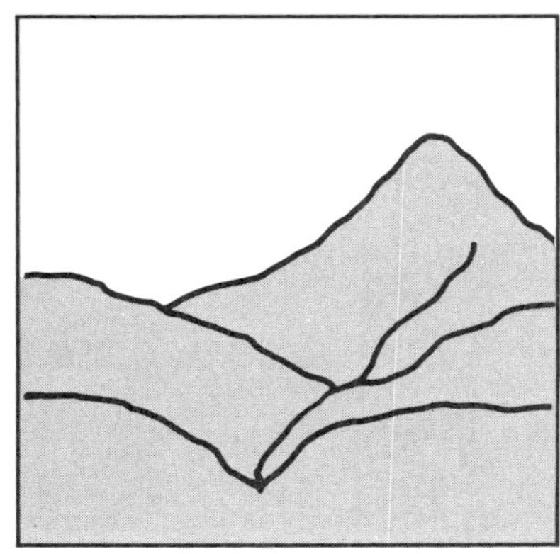

Shake, Rattle, and Roll

Name ______________________________

Many earthquakes have occurred in California. Most of them have done no damage. The information listed below shows the number of earthquakes that occurred in three California areas between the years 1980 and 1985. Plot this information on the graph provided. Use a different color for each area. Then, use the information to answer the questions below.

Number of earthquakes that measured greater than 3.0 in magnitude

	1980	1981	1982	1983	1984	1985
Parkfield, CA	104	102	91	89	84	103
San Francisco Peninsula, CA	67	43	66	71	69	59
Mammoth Lakes, CA	185	126	65	61	72	23

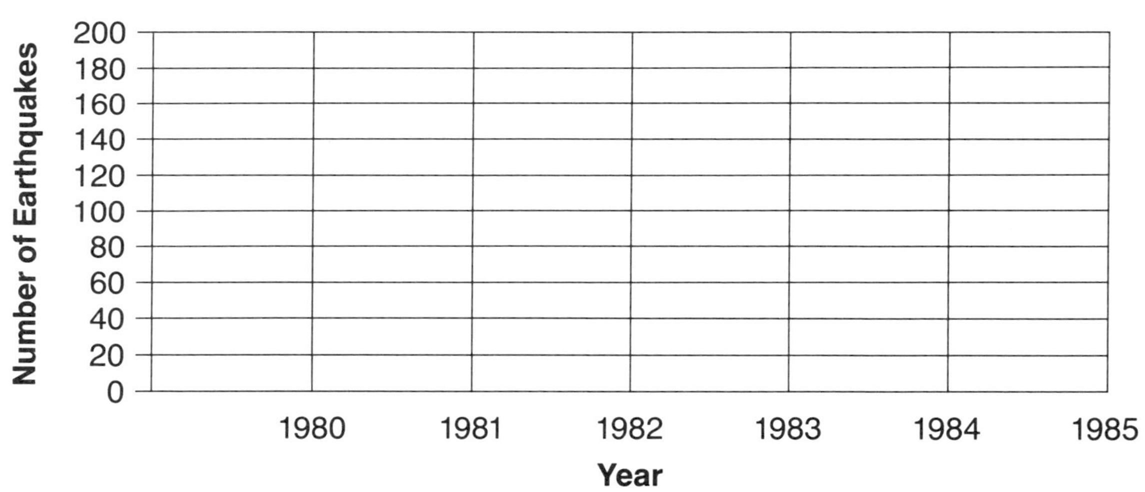

1. Which area had the most earthquakes from 1980 to 1982? ______________
2. Which area had the most earthquakes from 1982 to 1985? ______________
3. Which area had the fewest earthquakes in 1981? ______________
4. Which area had the fewest earthquakes in 1985? ______________
5. In which area did the number of earthquakes decrease most from 1980 to 1985? ______________

Hot Rocks!

Name ____

Some of the rocks we see come from volcanoes. There are different kinds of volcanic rock. The differences occur because the rocks are formed differently. Read the descriptions of the rocks below. Then decide how and where each rock is formed. Write the name of the rock on the appropriate blank.

Basalt: This type of rock forms when magma pours slowly out of the earth and cools slowly. Basalt has small crystals and sometimes has small holes.

Granite: This type of rock forms under the ground. The magma cools and hardens between layers of rock. Granite has large crystals and a coarse texture.

Obsidian: This type of rock forms when magma pours out of the earth slowly but cools quickly. It looks like smooth, black glass.

Pumice: This type of rock forms when foamy lava cools very quickly. It is full of holes, very light, and can float.

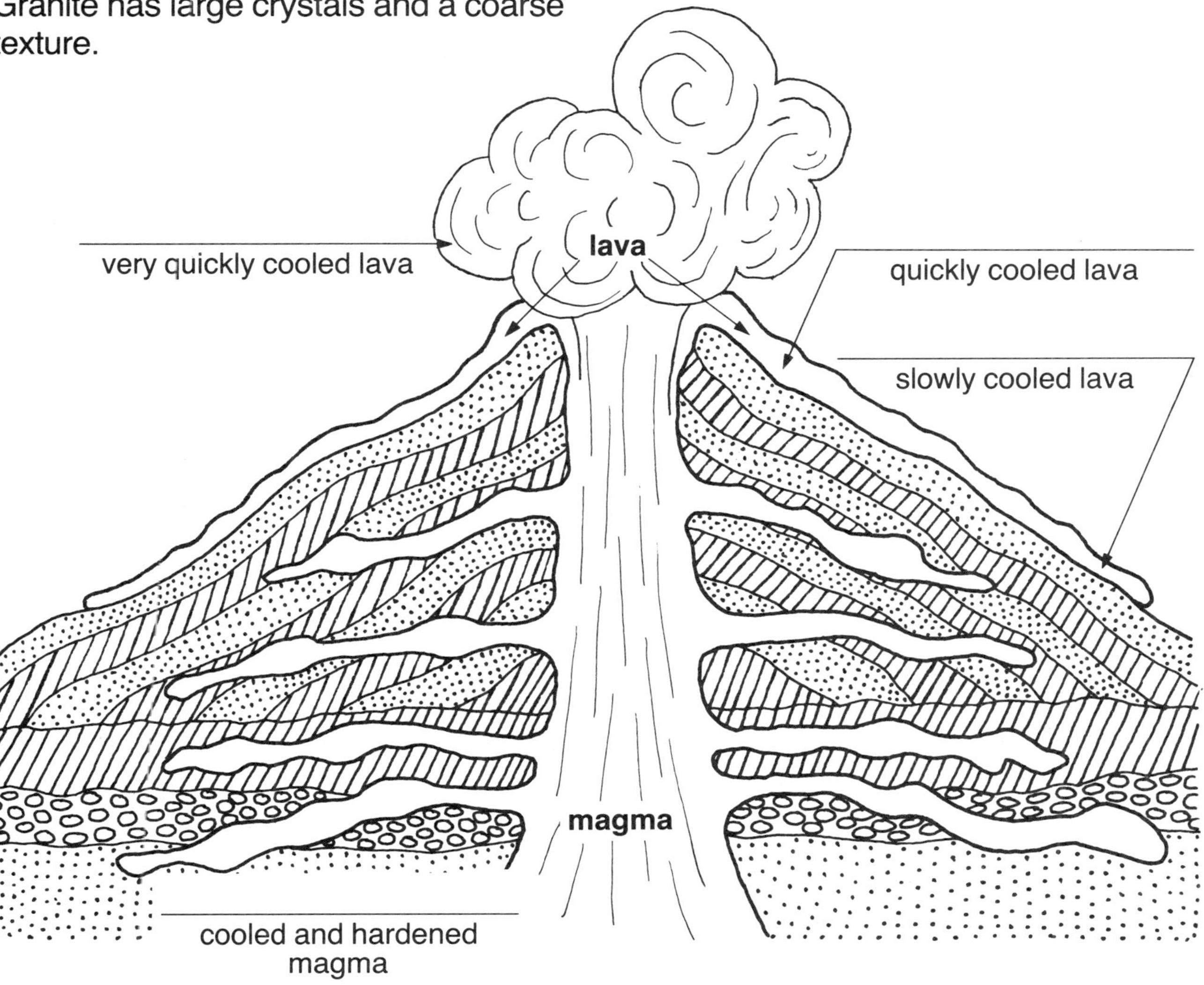

Don't Take It for Granite!

Name ____________________

Scientists often group rocks by the way they were formed. Read the descriptions of the three kinds of rock formation. Then look at the rocks below. Write the name of each rock in the column that describes its formation.

igneous rocks

Rocks formed by the cooling of hot liquid magma or lava.

____

____

____

____

sedimentary rocks

Rocks that were formed over a period of time in layers. They are made of the hardened sediments of broken rocks or other materials.

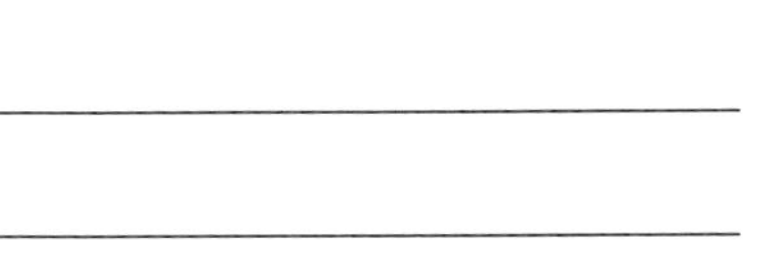

____

____

____

____

metamorphic rocks

Rocks that began as another kind of rock, but have changed because of pressure and/or heat.

____

____

____

____

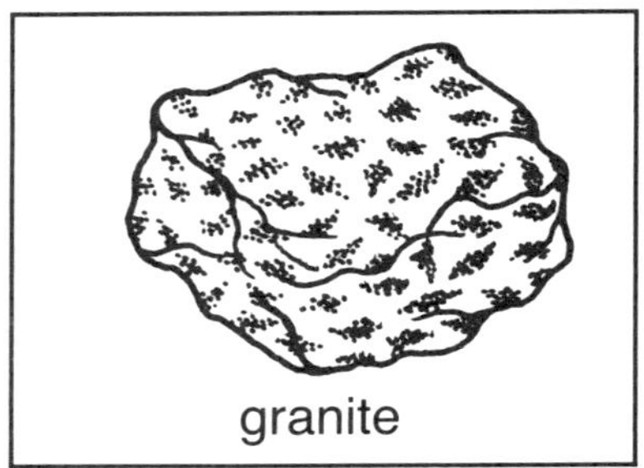

granite

Formed by cooling magma.

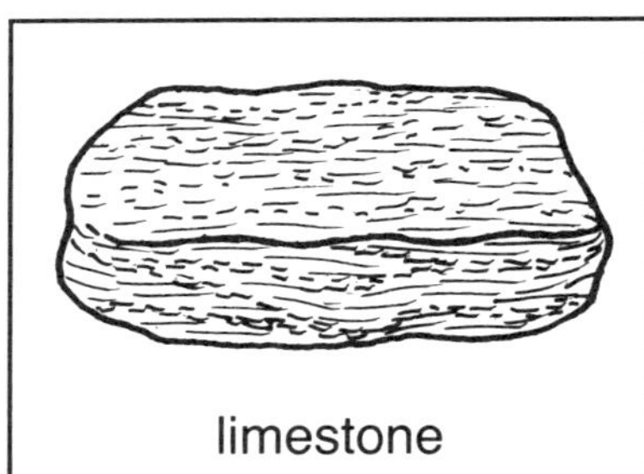

limestone

Formed underwater, made up of shells and skeletons of animals.

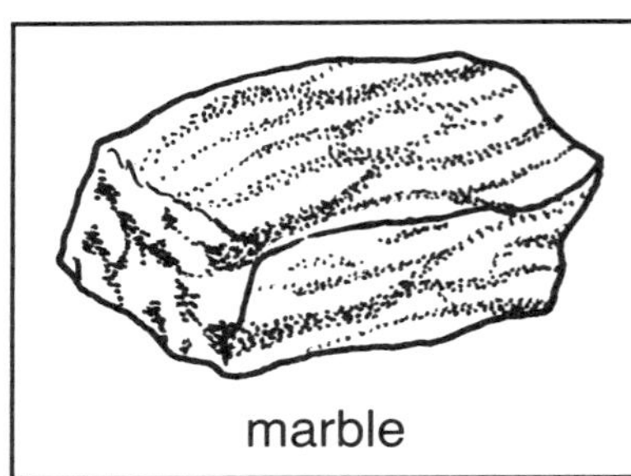

marble

Formed from limestone that has been exposed to heat and pressure.

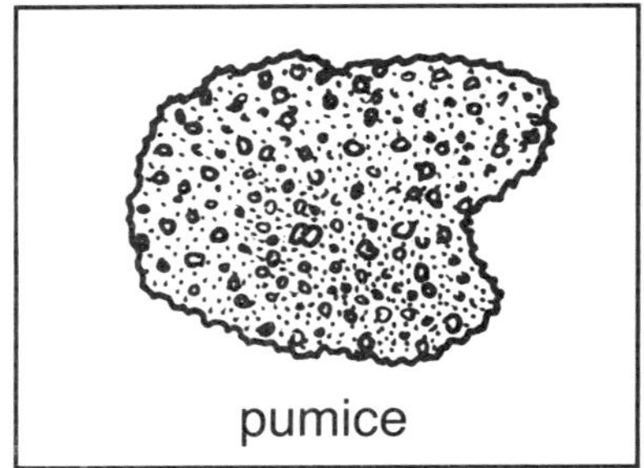

pumice

Formed by quickly cooled lava.

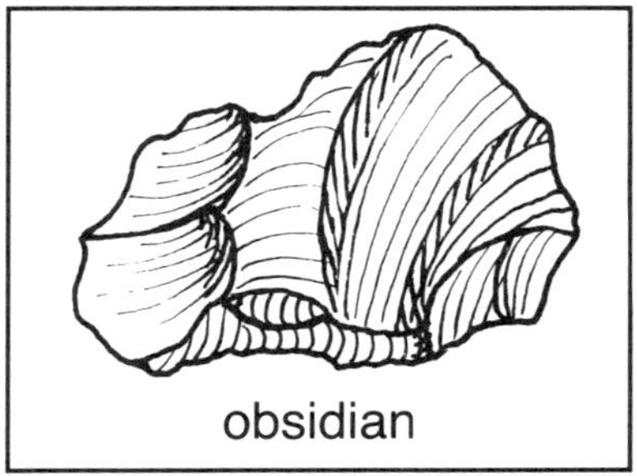

obsidian

Formed by slow-moving lava that has cooled quickly.

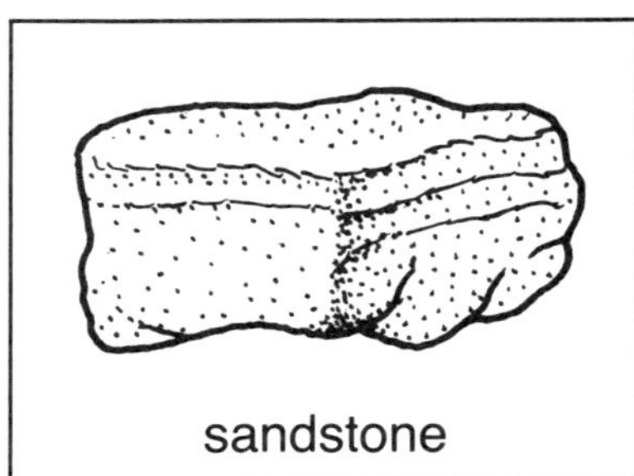

sandstone

Formed underwater when tiny grains of sand become cemented together.

Scratch and Match

Name ______________________________

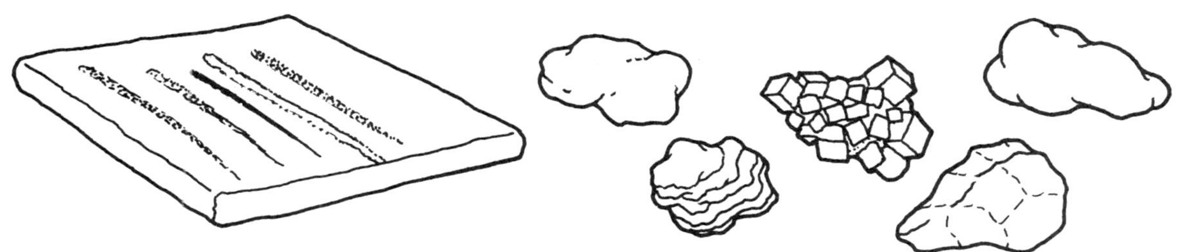

When you rub certain rocks across a piece of tile, they leave streaks of color. Scientists can sometimes identify the mineral in the rock by the color of the streak. Try this for yourself!

1. Rub a piece of rock across the tile. Record the color it leaves in the chart below. Then look at the color-sorting key and find the mineral the rock contains. Write the mineral name in the chart.
2. Repeat for the remaining rocks.

You will need:
An assortment of rocks
Piece of unglazed tile
Pen or pencil

Rock	Streak Color	Mineral
1		
2		
3		
4		
5		
6		
7		
8		

Color-Sorting Key

Color	Mineral	Color	Mineral
light blue	azurite	yellow-brown	limonite
greenish black	chalcopyrite	black	magnetite
bright red	cinnabar	green	malachite
lead gray	galena	green	olivine
red-brown	hematite	greenish black	pyrite

Telling Time with Fossils

Name ______________________________

Scientists can "see back in time" using fossils. Old fossils tell us something about the earth's past. Some fossils are more than two billion years old. Create a fossil time line. Look at the pictures below. Then write the name of each creature under its age on the time line.

Cephalaspis
≈410 million years old

Eohippus
≈50 million years old

Australopithecus
≈2 million years old

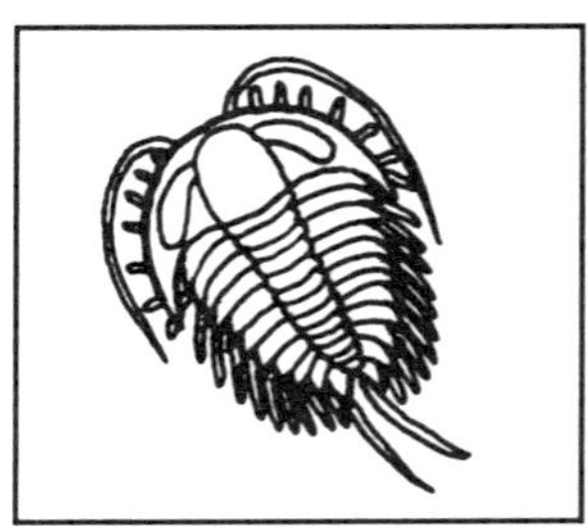

Trilobite
≈600 million years old

Dimetrodon
≈275 million years old

Stegasaurus
≈150 million years old

Ichthyostega
≈365 million years old

Diatryma
≈65 million years old

Millions of years ago

600 500 400 300 200 100 50 0

410 365 275 150 65 2

Fossil Circles

Name ____________________

There are many ways to group animals. This activity will show you one way.

1. Color the labels as marked.
2. Cut out the labels.
3. Listen for directions.

Color these labels red:

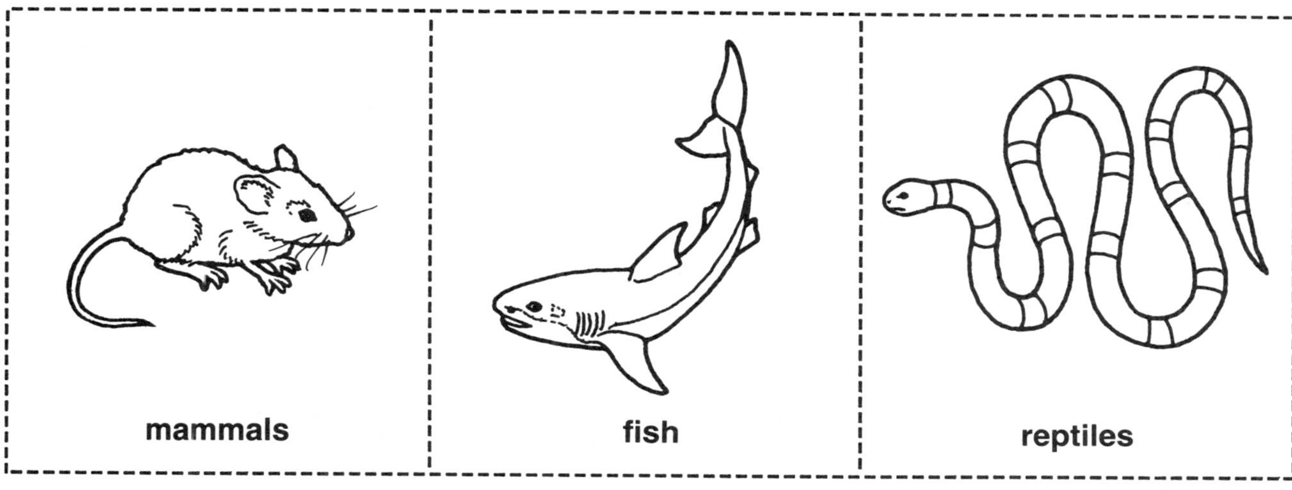

Color these labels yellow:

Paleozoic Era 600 million years ago to 230 million years ago	**Mesozoic Era** 230 million years ago to 63 million years ago	**Cenozoic Era** 63 million years ago to present time

Fossil Circles

Name ____________________

You will use these cards with the labels on page 35.

1. Cut out the cards.
2. Listen for directions.

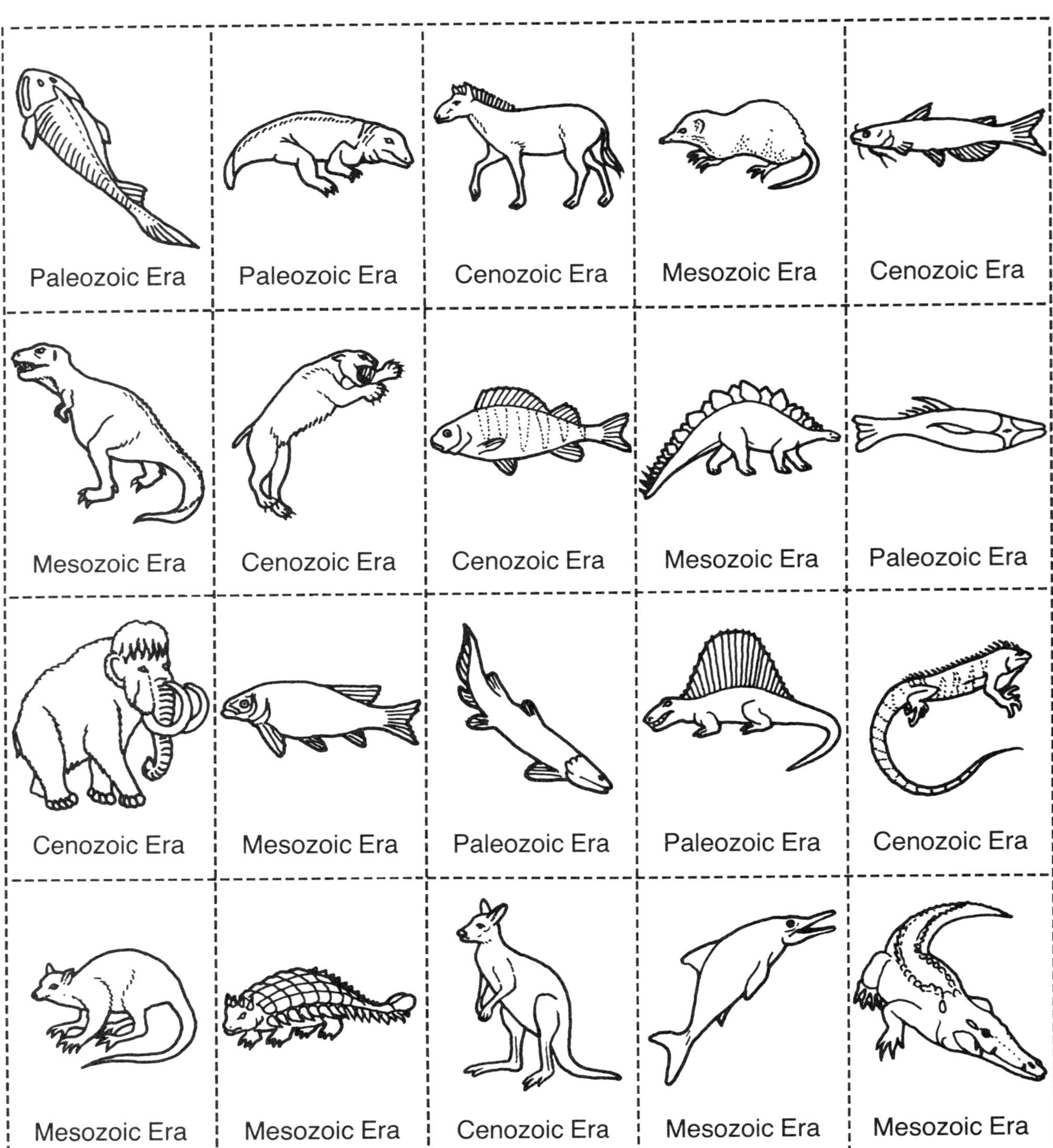

Inside or Out?

Name ____

Read the information about each planet. Decide whether it is an inner planet (less than 145,000,000 miles from the Sun) or an outer planet (more than 145,000,000 miles from the Sun). Write *inner* or *outer* on the blank below each planet.

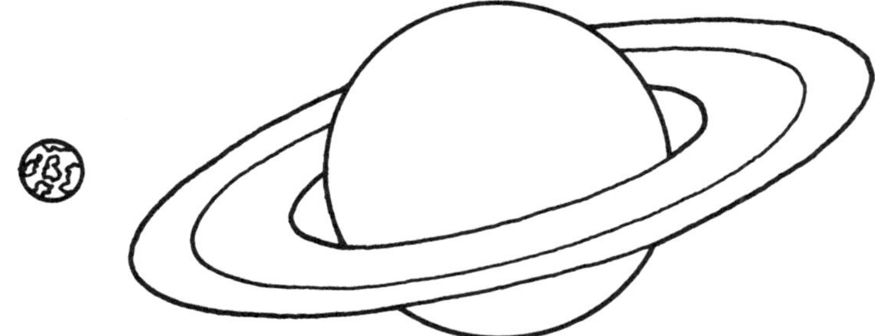

Jupiter
distance from Sun:
484,000,000 miles

____

Mars
distance from Sun:
142,000,000 miles

____

Saturn
distance from Sun:
886,000,000 miles

____

Uranus
distance from Sun:
1,780,000,000 miles

____

Earth
distance from Sun:
92,956,524 miles

____

Mercury
distance from Sun:
36,000,000 miles

____

Pluto
distance from Sun:
3,660,000,000 miles

____

Neptune
distance from Sun:
2,800,000,000 miles

____

Venus
distance from Sun:
67,200,000 miles

____

As the Worlds Turn

Name ______________________________

Each planet takes a different amount of time to make a complete circle around the Sun. For example, it takes about 365 days for Earth to circle the Sun. But it takes about 88 days for Mercury to circle the Sun, and 90,700 days for Pluto to circle the Sun. Use the following information to help you answer the questions below.

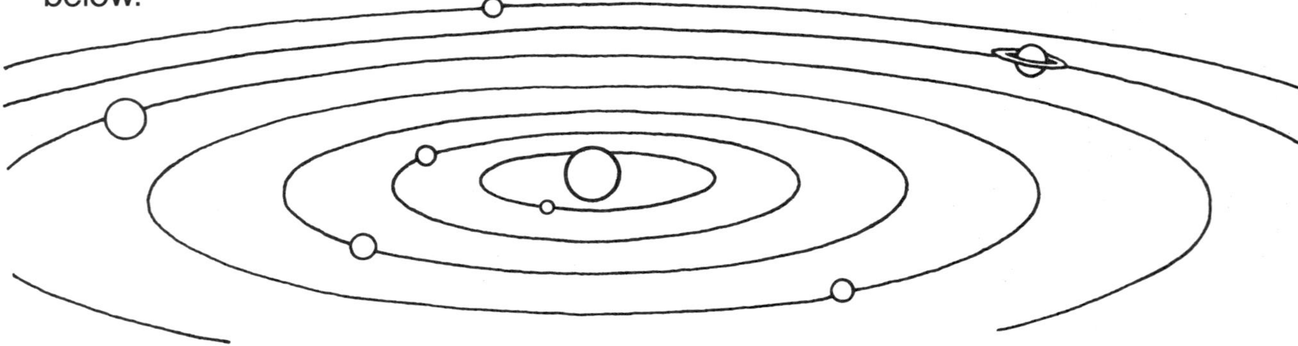

If you were this many Earth-years old: You would be this many years old if you lived on:

Earth	Mercury	Mars	Jupiter	Uranus
8	33	4	.6	.08
9	37	4.5	.7	.09
10	41	5	.8	.1
11	42	5.5	.9	.11
12	49	6	1.0	.12

1. If you were 8 Earth-years old, how many Mercury-years old would you be?

   ____

2. If you were 10 Earth-years old, how many Uranus-years old would you be?

   ____

3. If you were 12 Earth-years old, how many Mars-years old would you be?

   ____

4. If you were 41 Mercury-years old, how many Jupiter-years old would you be?

   ____

5. If you were 1 Jupiter-year old, how many Earth-years old would you be?

   ____

6. How many Mars-years old are you?

   ____

Faces of the Moon

Name ____

The pictures below show the positions of the moon over a period of a month. The names of the phases we see on earth are written next to each circle. In each circle, draw what the moon would look like from earth during that phase. (The first two are done for you.)

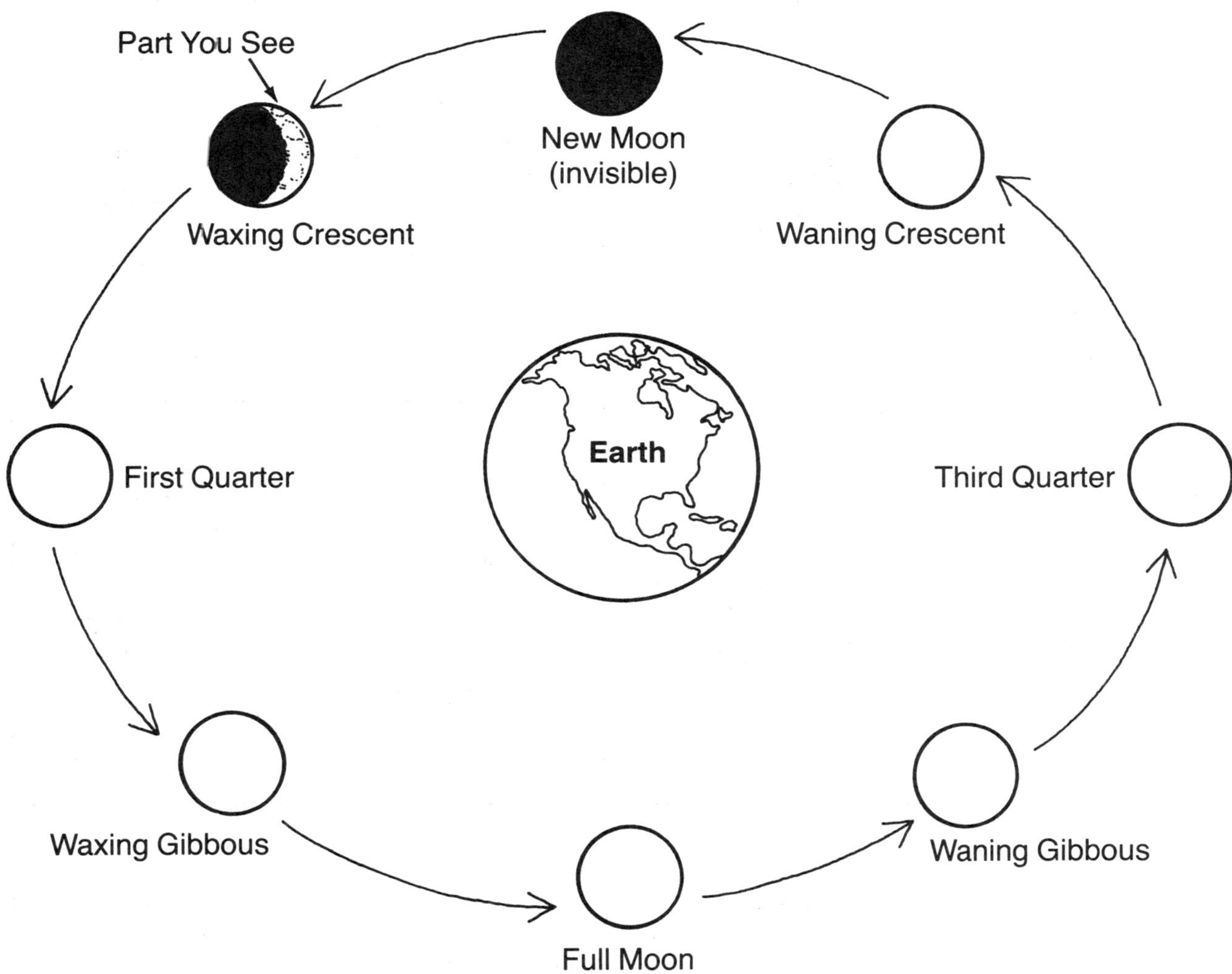

Eclipse It!

Name ______________________________

When the moon blocks the sun's light and casts a shadow on the earth, we see a *solar eclipse*. When the earth blocks the sun's light and casts a shadow on the moon, we see a *lunar eclipse*.

Circle the picture that shows the positions of the moon, earth, and sun during a solar eclipse.

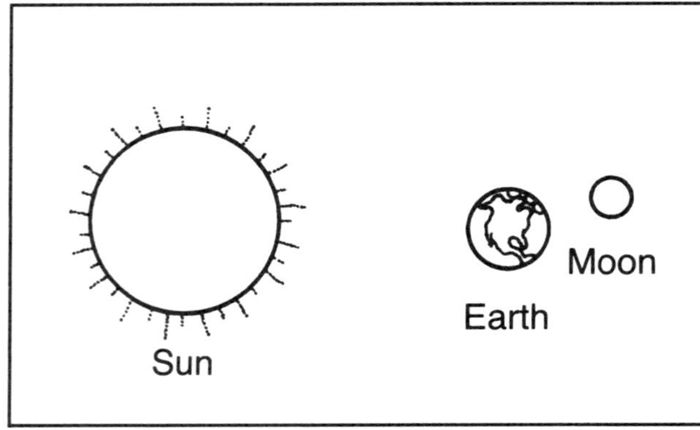

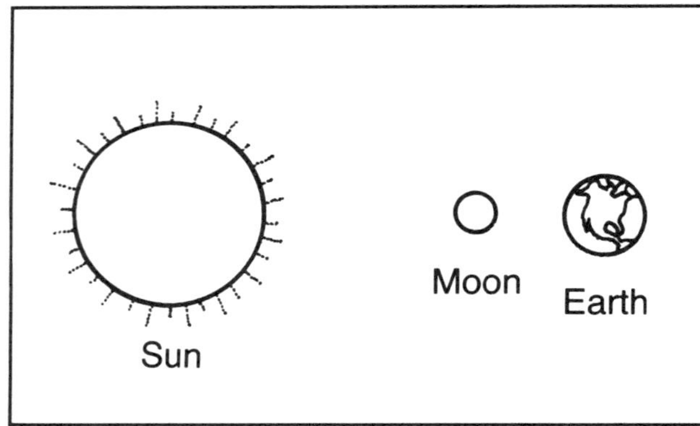

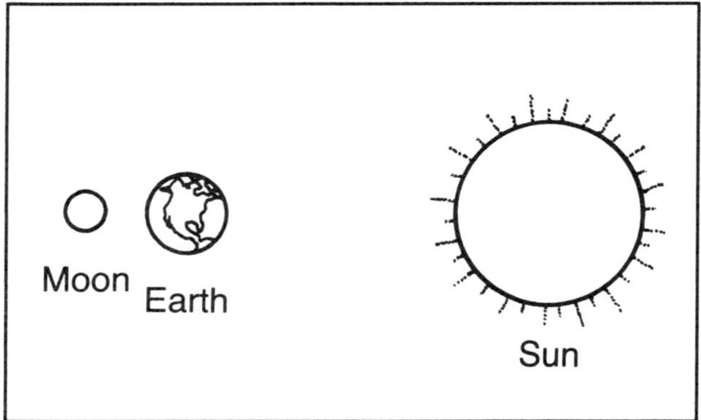

Circle the picture that shows the positions of the moon, earth, and sun during a lunar eclipse.

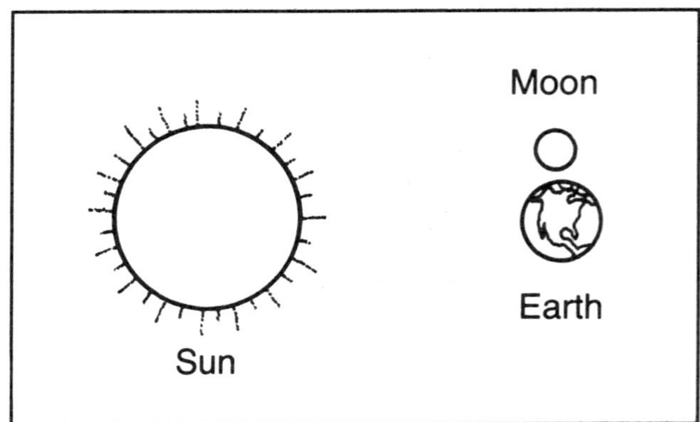

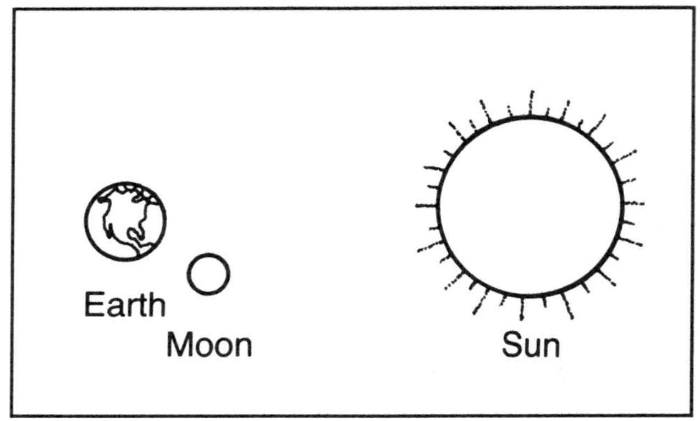

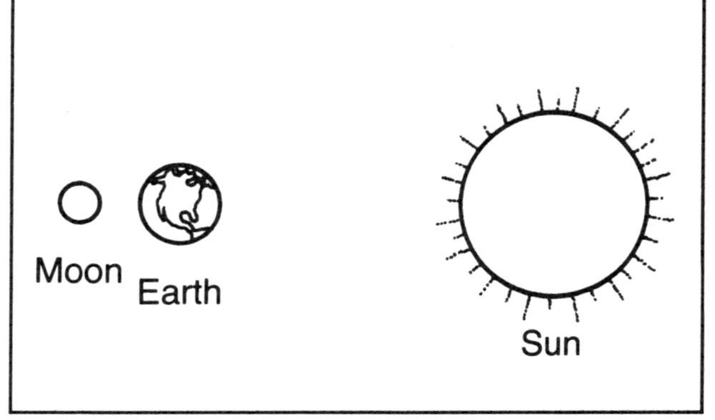

The Constellation Game

Name ____

Look at the stars in each circle. Can you see any pictures in the stars? When you do, draw lines to connect the stars. Give your constellation a name. Write it on the blank below the circle.

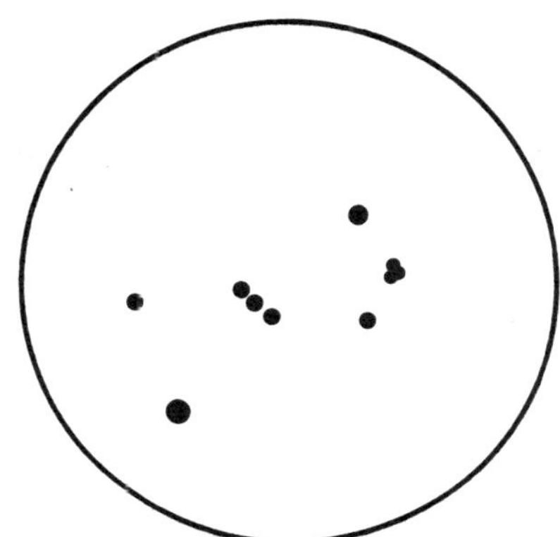

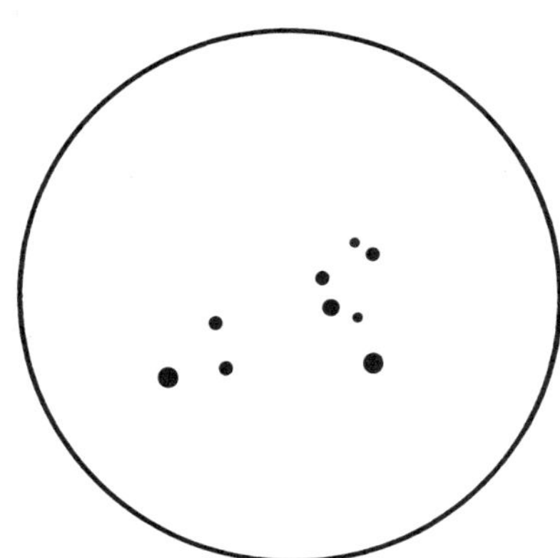

1. ______________________________ 2. ______________________________

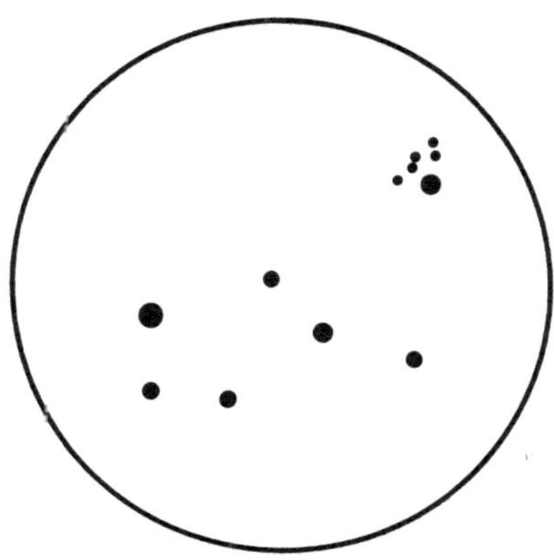

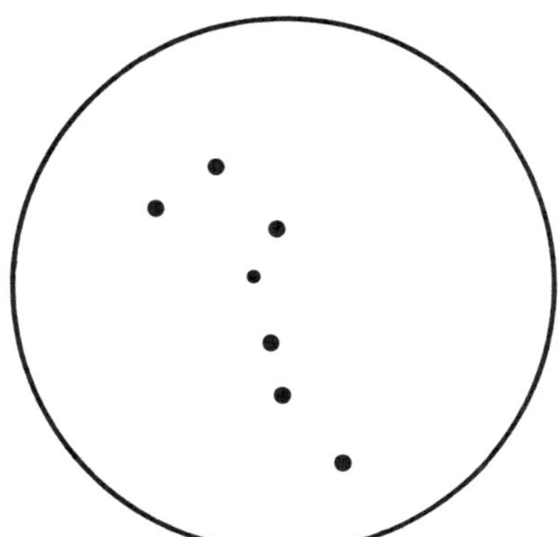

3. ______________________________ 4. ______________________________

Starry, Starry Sight

Name ____________________

The following are constellations we can see in the sky. Look at them closely and then try to find each constellation on the star map below. Draw lines to connect the stars in each constellation you find.

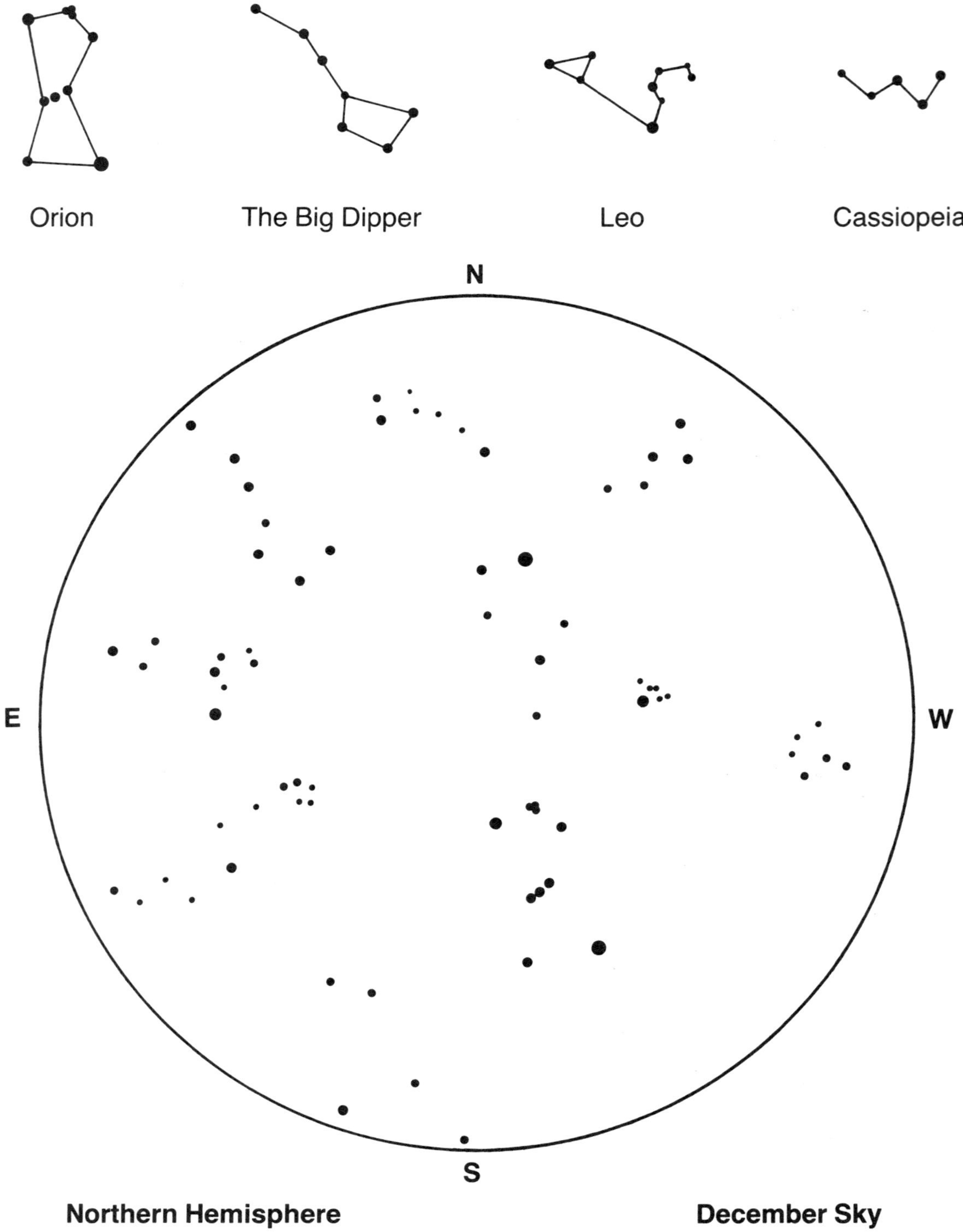

How Hot Is Red?

Name ______________________________

Scientists sometimes classify stars by their colors. Each color is associated with a temperature range. Use the following information to fill in the bar graph below.

- Blue stars have a temperature of 34,000°C or more.
- Yellow stars range in temperature from 5,000°C to 6,000°C.
- Blue-white stars range in temperature from 11,000°C to 34,000°C.
- Red stars have a temperature of 4,000°C or less.
- White stars range in temperature from 8,000°C to 11,000°C.
- Orange stars range in temperature from 4,000°C to 5,000°C.
- Yellow-white stars range in temperature from 6,000°C to 8,000°C.

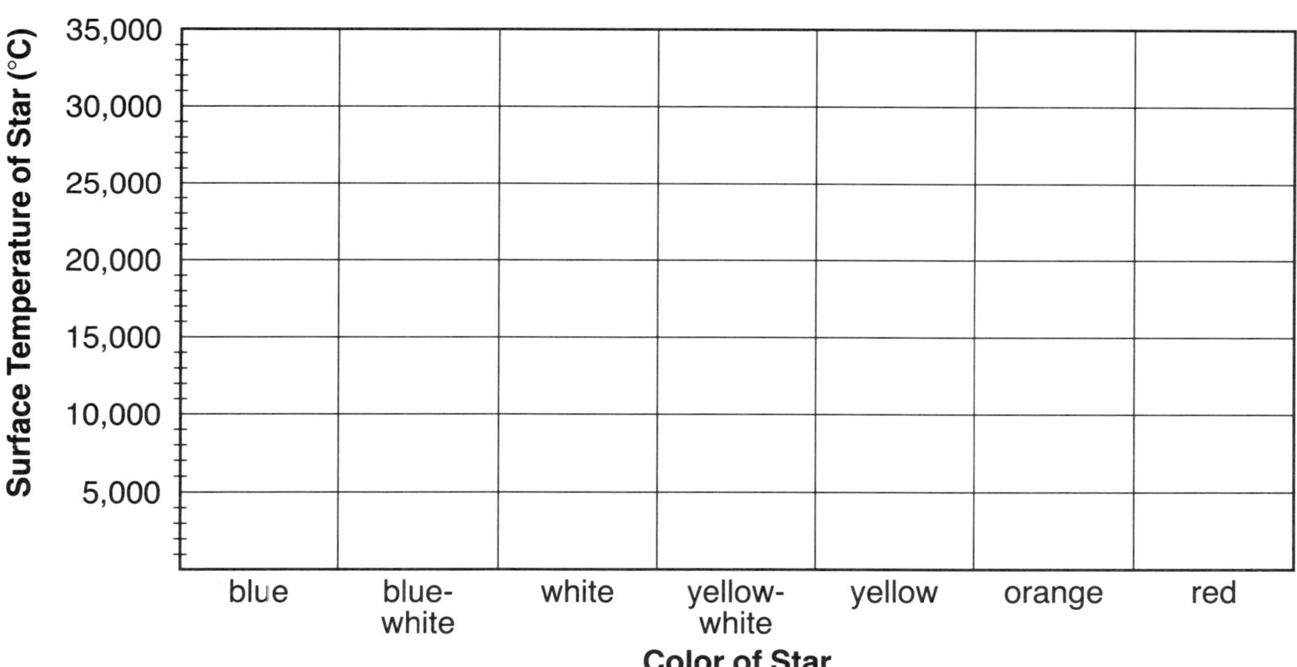

1. Which type of star is the coolest? ______________________________
2. Our sun is a yellow star. How hot is it? ______________________________
3. Sirius is 10,500°C. What color is it? ______________________________
4. Rigel is 12,000°C. What color is it? ______________________________